STRANGE CREATURES
SELDOM SEEN

STRANGE CREATURES SELDOM SEEN

GIANT BEAVERS, SASQUATCH,
MANIPOGOS, AND OTHER MYSTERY ANIMALS
IN MANITOBA AND BEYOND

JOHN WARMS

ILLUSTRATIONS BY JARMO SINISALO
AND NUMEROUS EYEWITNESSES

COACHWHIP PUBLICATIONS
Greenville, Ohio

Strange Creatures Seldom Seen, by John Warms
© 2015 John Warms

Cover Design by William Rebsamen

ISBN 1-61646-288-4
ISBN-13 978-1-61646-288-8

Coachwhip Publications / CoachwhipBooks.com

All Rights Reserved. No part of this publication may be reproduced, stored in a retrieval system or transmitted in any form or by any means—electronic, mechanical, photocopy, recording or any other—except for brief quotations in printed reviews, without the prior permission of the author or publisher.

CONTENTS

Acknowledgments	7
Foreword	9
Introduction	11
The Giant Beaver	15
Manipogo	54
Big Snakes	72
The Sasquatch	103
Big Birds and Big Bats	150
Beaver Ducks	165
Crocodiles and Lizards	168
Underwater Moose	174
Sea Dogs	184
Whales	188
Big Frogs	198
Dry Rats	203
Big Moose, Small Moose	205
Merbeings	208
Miscellaneous Creatures	211

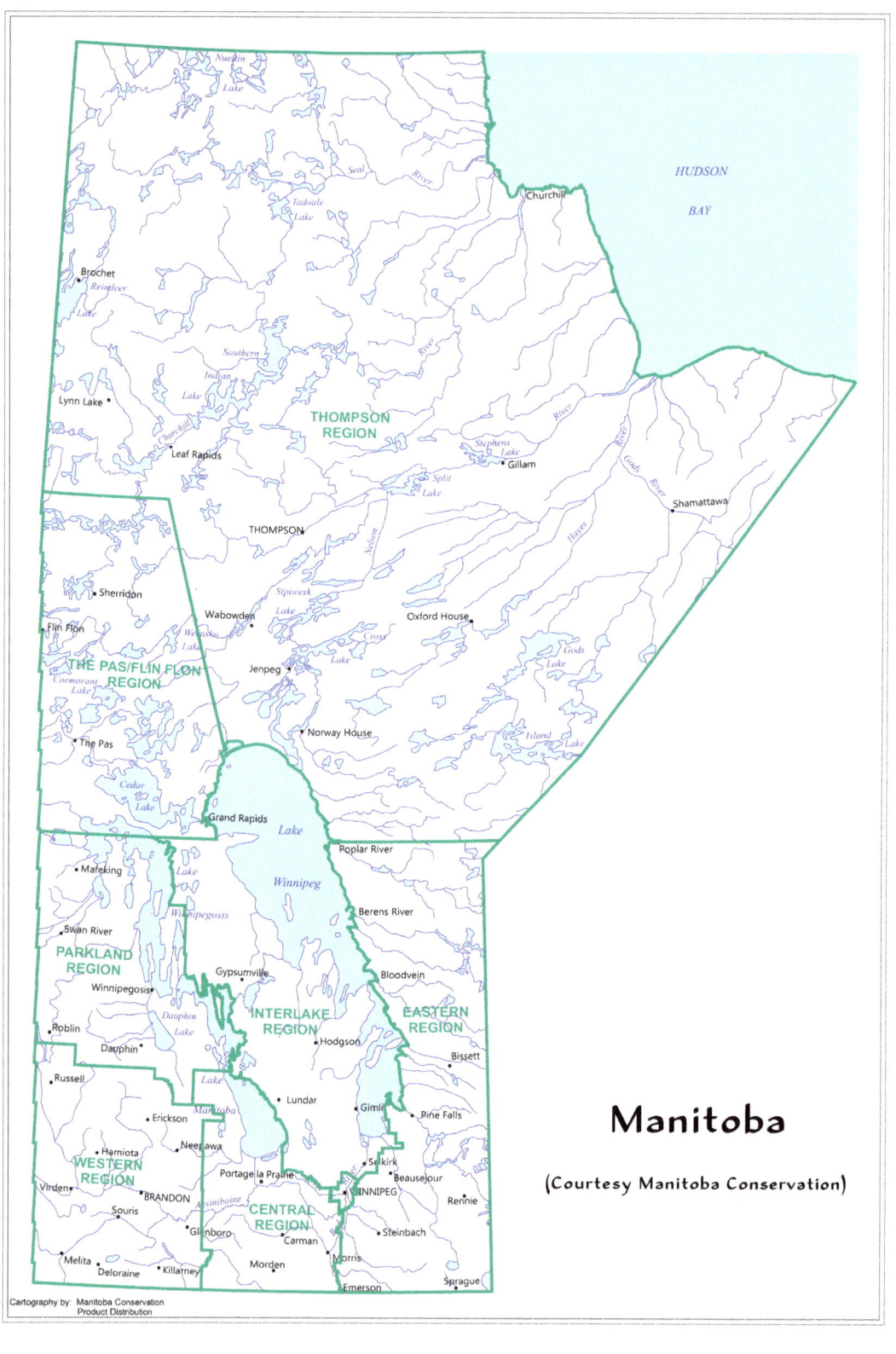

ACKNOWLEDGMENTS

WHEN I CONSIDER the contents of this book, I realize that I am overwhelmingly indebted to the hundreds of First Nations people, and others, who opened their hearts to me in blind trust, sharing their unique encounters with me, or relating information that was relevant to the topic of unusual animals. If I had not been able to draw upon their limitless wealth of knowledge, all this wisdom would just not be available. I also appreciate so much the many sketches that accompanied the stories, sketches that give such an enlightening reality to the creatures described.

I am also indebted to a number of professional people who kindly and generously contributed in some way towards my research, and ultimately therefore, towards the materialization of this book. One of my earliest contacts, as I searched for someone interested in doing DNA testing on a snake shed, was Dr. Brian Crother of Southeastern Louisiana University, who continues to be an advisor and ardent encourager along my way. Much closer to home, Dr. Robert Wrigley, curator of the Winnipeg Zoo when I first met him, patiently evaluated numerous sketches of unknown animals, and on many occasions helped me think things through. His offer to write the foreword to the book was a most pleasant and welcome surprise.

Some other professionals whose interest, advice, and encouragement I appreciated are Dr. Richard Harington of the Canadian Museum of Nature in Ottawa, the late Dr. Roy Mackal (University of Chicago biologist and cryptozoologist), cryptozoologist Gary Mangiacopra of Connecticut, Dr. Randy Mooi of the Manitoba Museum, and William Watkins of the Department of Natural Resources in Winnipeg. I cannot thank enough my artist friend, Jarmo Sinisalo, who so capably brought to life images of the various creatures that were originally only vague descriptions. I appreciate so much the participation of my young friend, Ken Reader, on numerous excursions, bringing with him the much-needed technical skills, advice, and enthusiasm. Without the keen interest, unflagging enthusiasm, and financial support of my sister, Connie, much of the research and subsequent book might still only be a dream rather than a reality. Chad Arment, a cryptozoological investigator from the U.S.A., is not only my publisher, but also a friend and advisor whose perspectives I value very much. And to all others who contributed in the form of advice, suggestions, proof reading, encouragement—it all means so much!

Thanks to my employment with the Office of the Vulnerable Persons' Commissioner, my work frequently took me to, or near, many remote communities where information regarding unusual creatures was available. These trips providentially not only legitimized my travels, but also enabled me to pursue them.

Many thanks also to friend, Susan Lackie, who not only assisted in our computer department, but also helped us cope with the modern electronics challenges.

It has been about two decades that I have been off on this inexplicable tangent, and I deeply appreciate my wife Sharon's kind tolerance of this passion which practically leaves her a widow for extended periods of time.

FOREWORD
Dr. Robert E. Wrigley, Biologist

SIGHTINGS OF STRANGE CREATURES have captured the imagination of people in all cultures, from ancient times to the present day. From mythical animals painted on cave walls, chipped onto stone slabs, sculpted in stone, clay and bronze, or the stuff of legends like Wisakedjak, our species seems unendingly fascinated by the unusual, the unexpectedly large, the creature that does not fit within our understanding of the natural and spiritual worlds. The sudden appearance of a strange creature often strikes confusion and fear in the observer—an instinctive response, and one with possible survival value.

Most individuals tend to keep an unsettling and weird sighting of a creature private to avoid ridicule from family and friends—akin to seeing an Unidentified Flying Object. It takes an understanding and keen enthusiast like John Warms to gain trust and to draw out information from a person who has seen a fantastic creature that he or she cannot explain.

As a scientist that has studied hundreds of public sightings of rare big animals, such as the cougar, it requires some real detective work based on limited data to sort out the probable real sightings from other potential animals, especially under certain conditions of lighting, weather, and possible mistaken scale of background vegetation. I admit to taking a close look at an unusual animal-like shape moving steadily on the surface of a river, only to conclude that it must have been a log in the current. But what about reports of running Sasquatch, giant beavers and lodges, and immense snakes slithering through the boreal forest? Many of the observers are highly experienced in the outdoors, and are familiar with the local wildlife. How does one explain these and other strange creatures described in John Warms' book? Is there more to this subject than over-active imaginations?

What interests me most about these startling observations is two-fold. It is fascinating to try to determine the actual identity of the creature (although not always possible), and then what made the observer conclude it was one of the beasts covered in this book. Have these individuals created descriptions and snap determinations based on prior stories or fears? Whatever the explanation, it makes experiencing Manitoba's wonderful outdoors all the more exciting. The author has made an outstanding effort to report honestly numerous strange-creature sightings in Manitoba. His narrative makes for a most unnatural history book, which I am sure many people will enjoy reading. They may even become more alert for that unexpected creature just around the next corner of the trail.

INTRODUCTION

I REALIZE THAT ONE GLANCE at the Table of Contents is enough to make anyone laugh. I did it myself—on numerous occasions! Anyone in his right mind would see the folly of believing in the impossible. How can a "moose" live under water? Or crocodiles and giant snakes survive in our climate? Or birds the size of airplanes go unnoticed? And merfolk? Mythology at its best!

How did I get tangled in this fantastic web of questionable phenomena? I, who had made it my life's goal to seek out the truth, and nothing but the truth. . . ? Considering the unusual subject matter contained within the covers of this book, it may sound preposterous for me to say that I have absolutely no interest in anything that is not true or real.

The delightful childhood stories of animals who spoke with one another are in the past—but they at least were creatures that were known, common to everyday life. What then was I to do with that first nagging rumor? I had two choices. Either I could ignore it and consider it nonsense, or I could investigate it—and then consider it nonsense. Or was there one more option? I chose the road less-traveled—and became, in time, a well-traveled fool!

I sought information from an eminent herpetologist who stated that Manitoba's largest snake was the garter. After I shared with him the rumor that a lady in Peguis had seen a serpent the size of a log in the middle of the highway, his shocking response included advice on the best type of bullet to use. When he noticed my reaction, he made a statement which will remain with me and influence my thinking the rest of my life: "If there's one, there's more."

Now, almost two decades later, the challenge of determining where the truth lay has taken me to the farthest reaches of not only my home province of Manitoba, but also to locations in the rest of Canada and the United States. I will never forget the exhilaration of my first trip into Manitoba's North where I quickly realized that this region was a mine of information waiting to be developed.

What initially seemed to be incredible stories, far from reality, slowly became credible accounts as they were duplicated time and again by strangers from unrelated communities. Where the first mention of a platypus was laughable, the second was sobering but exciting. Each subsequent story with corroborating details became a reassuring reminder that there had to be a common thread of truth linking them.

But merbeings? Was this not a bit much—like crossing a line? Certainly from a scientific standpoint they should not exist. But neither should the underwater moose . . . or water dogs . . . or jumping snakes . . . or huge burrows in sandy soil.

How we underestimate the marvels of creation!

I am not a qualified scientist, but just a reporter—dealing with stories from hundreds of witnesses who bared their souls and endangered their reputations in order to share a very significant event in their lives. In practically all cases, witness reports did not come my way at all. I was invariably forced to ferret them out after being given names of men, women, or children who were known by neighbors, friends, or relatives to possess information about a certain unusual creature. By far the majority of my informants were First Nations people, since they are found in communities near lakes, rivers and wilderness where much of the elusive faunae exists. Much of their knowledge goes back generations, and in some instances there are legends that have developed around them. I am convinced that some, if not all creatures that are addressed in the mythology, are, or were, based on real animals.

There are countless lakes across Canada and the United States that are rumored to contain monsters of some form or other. The lakes with especially deep areas seem to boast the larger species as a rule, but even some shallower waters have surprisingly large and varied specimens described as well. I have come to the conclusion that if a body of water or an area of land has been given a name depicting something fearful or mysterious, chances are that it or its environs are associated with a creature that had a significant impact on the indigenous peoples who named them.

Historical reports of dogs being offered to the lake gods in order to secure safe passage across big waters may be rooted more in wisdom than superstition. Perhaps the dogs served as a distraction (and a meal), allowing the canoeists to cross safely. From the many accounts of dogs disappearing from along shorelines, it would have been understood that some creatures considered them fair game.

There seemed to be consensus among elders across the land that danger lurked beneath their waters, indicative of the widespread warnings to children that they should not play in the water, or even near it. This is understandable, since the majority of unique creature reports are either of animals that lived entirely in the water, or both on land and in water.

Of the hundreds of contacts that I pursued in the last two decades, I can recall being rebuffed by only a single witness. I suppose the sincerity of my requests was obvious, and the topic of discussion unique. I sensed that there was often a feeling of relief or catharsis when I showed an interest in hearing about their sighting, and believed the story at hand—as if a burden were being lifted and shared with a friend. There were two other instances, however, that were of a negative sort. A young lady, overhearing my dialogue with her relative, told me emphatically that such hidden matters should be left alone and not discussed. The second episode involved a man who wanted me to leave his community immediately and never come back. A respected elder, after hearing of the incident, repeatedly assured me that this attitude was not representative of the majority in his community, and to pay no heed. Without the generous welcome and kindly tolerance of so many Native communities, I would not be enjoying the bountiful results of my quest, and be able to share them with the world.

Did I believe every story I was told? Since the witnesses did not approach me, that of itself was a positive aspect to begin with. Their sincerity was readily determined in their demeanor, so I generally had no misgivings about the information they shared with me. Furthermore, unless the topic was a species of animal that I had never heard of before, there was usually a corroboration of details from earlier witnesses that was not only reassuring but exhilarating. There may be the odd account that is exaggerated or contrived, but on the whole, I feel confident that by far the majority of stories were given to me in good faith—rational assertions of ordinary folks who were going about their daily activities.

Frequently witnesses would state that they didn't talk about their encounters because people tended to make fun of such stories, so I knew that they had much more to lose than to gain in revealing their secrets. It was common for me to be informed, "You are the only one I've told this to except for my husband (or wife). . . ."

Although the Native Peoples are the most likely to be privy to information about these largely unknown creatures, non-Natives are joining their ranks in increasing numbers due to the prevalence of northern travel and residency.

I feel immensely grateful for the privilege of laying eyes on two of the more popular cryptids in this book—creatures that I deliberately pursued because of information received regarding their habits and habitat. The first one I saw—the Giant Beaver, which is believed to be extinct for over 10,000 years—was exactly where I had been told it lived, and the same with the second one—the Giant Snake—which surfaced shortly after the deployment of a luring technique simulating activities that were known to attract them. In future, therefore, if we go to places where sightings have occurred, there is no reason why new creatures cannot be documented on a continuing basis.

Except for those who were traumatized by their sighting and wished they had been spared the encounter, most witnesses seemed reassured when they were told that they were not the only ones with such an experience, and the desire to know more about their particular creature was obvious. "When will your book be finished?" was a frequently asked question. I might add that the publishing of this book has been delayed for years simply because of so much "privileged information" that it contains. I can understand why the location of certain significant sites was not shared with me. There is a perception of sacredness surrounding mystery creatures and their habitat—and one is loath to invade the privacy which they enjoy.

I never dreamed that my search would take me far beyond the world of giant snakes into the seemingly endless parade of unusual creatures that you will be introduced to in this book—and, an introduction is all that it will be, since so little detail is available, to date, on most species. Perhaps this publication will encourage others to come forward with their stories, and hopefully the halls of learning which shunned new information will yet embrace reality and join in the endless research which is inevitable. The people have spoken, and we will do well to listen to them. I find their individual and collective voices unanimous and credible. Fortunately, truth does not change, regardless whether or not it is believed. Only time will vindicate those who dared to speak of what they saw, and then they will be appreciated for allowing themselves to be vilified for a season. I suppose each creature will likewise be subjected to the ultimate reality test, and, until it passes, it will remain suspect.

Contrary to conventional wisdom where seeing is believing, I found myself rewarded on two occasions with seeing because of first believing. I hope we can do much more of the same.

THE GIANT BEAVER

I'M SURE SOME FUR TRADER shook his head in amazement on seeing the size of such a beaver pelt, and paid well for it. The manufacturer who turned it into felt hats would have been equally astounded at the black bear-sized fur. All that saw it, including the trapper, would have thought the same thing: grandpa of all grandpas!

Similar enough to the ordinary beaver to be taken simply as "a big one," it got passed off as no more than an anomaly. Had this been a frequent scenario, however, it would most likely have influenced our awareness of the creature. As we'll see later, there are obvious reasons why the people involved in the fur trade, other than the trappers, seldom or never saw a huge pelt.

Today we know from recovered bones that the giant beaver is in a separate classification from the one we know so well as *Castor canadensis*. It has been named *Castoroides ohioensis* after the state where its remains were first documented, and there is evidence that its habitat stretched from Alaska to Florida.

It is also widely held that this big beaver has been extinct for over 10,000 years. But, this is where I beg to differ.

Let me explain, and take you down the long road I traveled.

In my search for stories of unusual creatures, I was also told about a variety of other things that people had seen or heard—things like lights in the sky, objects in the sky, holes in the ground, tunnels, caves, strange sounds, odors. . . .

When I finally investigated the tunnel stories, I was pleasantly surprised. They all matched. They were the same size, the same shape, in similar landscapes.

No one seemed to know what made them.

Running water was ruled out. The elliptical shape was consistent with what a burrowing animal would make, and the "table-top-smooth" interior suggested something that had brushed against it innumerable times.

What creature would fit into a burrow about three feet wide and almost as high? What animal would be capable of making it?

The author and a life-sized replica of a giant beaver at the Beringia Museum in Whitehorse, Yukon.

The answer became evident as soon as I began combing the north for information. At South Indian Lake, when I enquired about the presence of any unusual creatures, the big beaver was the first one that was mentioned—a black bear-sized beaver.

"How big are their houses?" I asked naively, and as soon as the words were out of my mouth, I knew it was a foolish question. Or was it? If their lodges were proportionately larger than those of ordinary beavers, there would be no mystery. Their huge houses would have been noticed long ago—or so I thought.

The response came as a surprise: "Tunnels. In sandy banks and islands."

When the information finally sank in, I realized I had the answer to the question of what had made the tunnels discovered in southern Manitoba.

Where are these tunnels? Obviously, they would be in the high banks of lakes and rivers, and indeed many of the known sites are located along the shores of postglacial Lake Agassiz and its tributaries. Just as our typical beavers burrow into banks where necessary, the giant beavers seem to do likewise in most instances.

But, is the opposite true as well—do the big beavers ever build a house of sticks and mud? An emphatic "You will never find a house they made!" was the ironic statement from an old trapper from South Indian Lake, even though some of his own people told me otherwise.

From the ever-increasing number of stories I am hearing, there are definitely some large lodges and dams scattered throughout the north, with the building materials consisting of logs rather than sticks, and stones bigger than what humans could carry. These large houses and dams seem to be confined mostly to isolated areas in the Canadian Shield, presumably due to the scarcity of ideal sandy banks which are more plentiful in the south.

Even when large beaver lodges were encountered by trappers, it seems they were not fully aware that the residents were not ordinary beavers. For one thing, the creatures were not often spotted. As someone quoted from the elders who spoke about them, "They don't come out often."

I noticed a developing pattern after I spoke to a number of trappers. In certain oversized beaver houses where traps or snares were destroyed, they would invariably mention that they had caught some adult-sized beavers there. They deduced that the ones that continually broke their traps must have been considerably larger and more powerful than the ones that didn't get away. Furthermore, the more astute observers noticed that the ones they did catch in these unusual houses appeared to be juveniles, even though they were adult size, because of the softer, less muscular body structure. Most of these trappers noted that the pelts were also darker than those of normal beavers, usually describing them as black.

One trapper from Pukatawagan told me that, after catching four super-adult-sized beavers in one lodge, he sensed that something did not add up. He said he normally would catch only two adults per house, and any others caught would usually be smaller because they were from a younger generation. He also noted that the beaver house "was as big as *my* house."

When he went back to this lodge the following year, it was vacant, and the water level around it was down enough that he could see the three foot diameter entrance, about three times the size of normal openings. He knew that no injured beavers had died in the house or "a bear would've ripped it apart to get at the meat." Either the adults that broke the traps died somewhere else, or they got away and abandoned that lodge. Therefore the dam fell into disrepair, allowing the water to drain away and expose most of the house.

It seems that whoever tried to trap or snare these giant rodents found out the hard way that they were not to be messed with, leading most hunters to advise their descendants to avoid the area altogether. I found a few people, however, who did encounter them in broad daylight, and their observations are proving to be invaluable additions to the limited knowledge we have about them.

It is fitting to begin with the first eyewitness account ever told to me regarding a present day sighting of the giant beaver. This was in the very community where this creature was first brought to my attention.

A young man was traveling by snowmobile near South Indian Lake one spring in the mid-1990s when he noticed a large, dark object on the ice. Leslie stopped several times to take a better look, wondering if it were perhaps a rock that he had not noticed before. When he got nearer, however, the object took several steps forward and disappeared into the water.

From the same community I got another account that closely matched the first one, only in this case the animal was sitting on the shore. This was in May, when there was still some ice on the lake. In the spring of 2006, a goose hunter from South Indian was following the wooded shore of a lake when he was shocked to see a huge beaver in front of him. Its size was so intimidating that he didn't dare challenge it with his shotgun, so he slowly retreated and watched as it went down the bank into the water. (When Bernie showed me the exact spot a few months later, there was no evidence at all of any activity, so it seems that quiet meditation is their sole purpose on land or ice—except where they need to construct a lodge or dam.)

A lady from Nelson House tells this story:

She and another woman were picking cranberries on a steep slope beside a lake when something suddenly jumped over them from above and hastened to the water below in a clumsy, waddling fashion. At about the same time, another bear-sized creature skirted where they stood frozen, and also made for the shore and disappeared into the water. Berry picking was over, and not just for that day!

If it had been common black bears, they would have been recognized as such, and not have disappeared underwater, since bears swim with their heads above the surface. The lady and her husband commented that the old people used to talk of seeing big beavers in certain waters, but efforts to catch them usually failed.

A man from this same community told me that he had seen a bear-sized beaver jump into the water at Apegano Lake, west of Paint Lake, when their helicopter approached. Several other witnesses also noted that these creatures are in the habit of "plopping" into the water rather than entering it gently as most animals do.

Many a trapper apparently lost his traps or snares to something that was considerably bigger and more powerful than the average beaver. The odd one that didn't get away was shot or caught in multiple snares, with the following story out of Pukatawagan being an exception:

Two friends were checking their traps together when they discovered that their snare, and anchors, were all missing. One of the anchors that the snare had been tied to was a large rock that should have been more than sufficient to hold an ordinary beaver. Another anchor, a log, was spotted down river, held in place by something under water. When they hauled it ashore, to their amazement a dead bear-sized beaver came with it, a tight snare around its neck. Its mighty strength had enabled it to get away, but only until it drowned due to the heavy anchor.

Since it was much too large to put into their canoe, which already contained some beavers, they skinned it on shore. The tail was about ten inches wide, and the body the size of a medium bear. Nearby was an inconspicuous structure, plastered against the side of the bank, visible only in summer.

Angus, the only surviving partner, estimated that the size of the pelt, adding length and width together, would have been about 140 inches, where an ordinary large hide could reach 70, and on occasion, perhaps 100.

In South Indian Lake lies Sand Island, known to be the site of many beaver tunnels. Apparently there is indication that they have recently collapsed, and the higher lake levels related to the power dams are identified as a factor. But before this came about, a couple had camped there and reported hearing noises below them, and felt a movement of the ground.

A similar phenomenon is mentioned at Nelson House. Before the flooding took place there, families lived here and there, some on the mainland, and some on islands. After putting up with the subterranean noises and the sensation of something moving around below them, the resident families eventually moved off Tait's Island, leaving it to whatever was making its home there.

A more recent visitor to this island told me that he heard a droning sound there once when he shut off his power saw, "like a diesel engine running." He walked around the small, sandy island in search of the noise, but kept hearing it wherever he went. Finally, after almost an hour, it suddenly stopped, leaving him still bewildered.

Another man from Nelson House described his experience from about 50 years ago when he had come upon an unusually large beaver lodge that was surrounded by stumps that were as tall as he. Something seemed to notice his presence and came swimming out of the house, leaving a trail of bubbles as it went back in and then out again without breaking the surface. What especially impressed him was the shaking of the ground where he stood when it went in and out of its house, as if it clumsily bumped into parts of it coming and going.

His uncle had apparently experienced the creatures as well, losing traps and snares to them. He had also watched them swimming, sometimes with their big heads above the water. He thought they were carrying branches in their mouths until he realized that it was their huge whiskers that he saw, the same conclusion that had been reached by a couple who saw a similar sight in the moonbeams as they sat in the Nelson House Nursing Station not many years ago.

These creatures that are seen now and then by the folks of Nelson House, swimming in daylight or by moonlight, may well be giant beavers which persist in living inconspicuously among them, deep in the high sandy banks of the community.

Back to the topic of the earth moving—an acquaintance related a recent experience on a hunting trip north of Cranberry Portage. As he and his partner were walking on a river bank, they were both shocked to feel a significant bumping sensation beneath their feet.

Two brothers from Pukatawagan out camping beside a distant creek reported some additional aspects to their experience. Not only did they feel the ground move below them, but they also heard whistling sounds, and had their gear soaked by something that shot or splashed water over them in the dark. Three times water fell on them from above, forcing them to move their bedding farther back from the water's edge. In the morning when they checked out the creek, they discovered some bubbles coming from a large hole a few feet from shore, so deep that a long stick could not fathom it. When I asked what creature might have made it, Miles, one of the witnesses with whom I have discussed the incident off and on for almost a decade, said he did not know.

In the fall of 2008, a friend and I had the opportunity of flying over that area in a helicopter, looking for reported foul-smelling caves, but we also examined the spot where Miles had made an X on our map. Sure enough, near the edge of the creek, a large hole was clearly visible under water. Its diameter looked to be in the neighbourhood of eight to ten feet, its perimeter built up much like the edges of a little ant hill. It appeared that whatever made the tunnel had deposited some of the material all around the opening.

As we flew away, I looked for one feature of the land that might hold a clue to the identity of the denizens of this site, and I was not disappointed. Although the area was covered with smooth rock hills typical of the Canadian Shield, the north hillside directly opposite the creek appeared to consist of soil reaching right to the top of the hill where solid rock became visible. Because of the proximity of the hole to this bank of soil, it is my belief at this time, given all the evidence, that *Castoroides ohioensis* lives there. If it were not for the fact that the site is so extremely remote, it would be an ideal spot for further study. However, since there are other sites much closer to home that have also yielded some evidence, they will be given priority.

One old man from Pukatawagan whose father had shot a bear-sized beaver in the head, helped him skin and stretch the pelt. He told me it was the size of a moose hide. The large lodge nearby he just took to be a multi-family dwelling.

Another witness compared the shape of a large beaver lodge to an igloo, as it appeared to have an extension built to it.

A lady living along the Alaska Highway witnessed what she at first thought was a black bear in the creek down a steep bank behind her place. It splashed water up with its paws, did a summersault, came up, and splashed some more. Then it dove underwater and disappeared. We agreed that this was not typical bear activity.

From a Sioux community in southern Manitoba come accounts of big beaver encounters experienced by ancestors who lived in the north-central States in the 1800s. One old man I spoke with said his grandfather helped kill a creature that was constantly damaging their fish trap on the Missouri River, and it turned out to be a giant beaver.

It would appear from this account, and the following one, that giant beavers may include fish in their diet.

A one-legged veteran of WWI made his living by fishing in the Assiniboine River which flowed by the Sioux Valley Reservation where he lived. On one occasion something grabbed his oar and nearly capsized his boat, but its teeth gradually lost grip on the blade, leaving four distinct tooth marks—two on each side. Then it swam under the boat, nearly tipping it again. This gentleman had complained about beavers spoiling his nets.

The narrator of this account remembered hearing his grandfather's friends sharing stories about a creature with "big whiskers and big eyes" living in the water, a creature he understood to be a huge beaver. People were discouraged from swimming there!

In this same community, tunnels matching the description of others discovered in southern Manitoba were exposed some years ago when a backhoe dug garbage pits in people's back yards, but no one knew what made them. Everyone suspected that they were still in use, so the voids were promptly filled in. However, one rumor indicated that garbage thrown into one pit had been disturbed by something crossing the gap in the broken tunnel, reinforcing the belief that something was still living there.

According to legend, tunnels run about 100 miles from the Assiniboine River in the Portage la Prairie area, north along the ridge created by Lake Agassiz on the western shore of present day Lake Manitoba as far as the Narrows. Big snakes were thought to inhabit these tunnels, and perhaps in some instances they did.

As accounts from witnesses who had seen the giant beaver or their tunnels increased, I became anxious to see for myself the phenomena being described to me. So, in the spring of 2006, convinced that the giant beavers still lived in the Assiniboine River, I made my first attempt to pursue the creature, and returned to the Sioux Valley First Nation where I had heard of the exposed tunnels the fall before. The result of that effortless experience was stunning, considering that this was my initial foray, with virtually no expectations. I had business in the area in mid-April, and, the weather being unseasonably mild, I decided to spend a night beside the river.

I arrived about an hour before sundown, and discovered the river in flood stage, with the water almost to the top of the present-day bank. I chose a spot across the river from where a gentleman from the community had camped for most of the previous summer. I had not yet had an opportunity to meet him, but from different sources I gathered that he had seen something, and references to a big hole in the bank matched my expectations. Had I talked to him before that day, I would have been disappointed, but as it was, I felt hopeful that I was standing beside a river containing the giant beaver.

All the stories I had heard went through my mind as I wandered among the trees beside the swollen river. I pondered them, and wondered why knowledge of the big beaver here was not widespread. There were only a few reports of brief sightings of something unfamiliar in the river, so it was definitely not general knowledge that a bear-sized creature with big eyes and long whiskers lived there. This dearth of information indicated a creature whose presence and habits were almost a secret.

The people at South Indian Lake had given me the same impression. Most of their sightings were in the spring, when trappers and goose hunters roamed about. It seems that the long winters which confined the beavers underground and under ice prompted them to escape briefly as soon as pockets of open water appeared.

I wondered about the alleged hole in the opposite bank, now deep below the surface. I visualized huge underwater tunnels that led across half a mile of river flat to the high, sandy banks where they branched into a network of burrows, offering shelter and whatever else the creatures made them for. Local tradition maintained that a tunnel went under every house!

I recalled the first tunnel that had ever been brought to my attention. Ed Grift, a farmer from Swan Lake in southwestern Manitoba came to my campground in the north Interlake almost every summer, and each time he would mention the tunnel he encountered in 1996 when he had a road grader come and level off part of a hill in his farmyard. Ed said that the grader wheels fell into a large burrow of some sort exactly where he hoped to erect a hay shelter, so he had flattened the tunnel for about 50 feet with his front-end loader.

When I finally had occasion to travel in that direction years later, I stopped at the Grift house. As we sat at his kitchen table, Ed reviewed the story again, and pointed out that the interior of the tunnel was as smooth as the table top before us. It had measured about three feet across, and almost as high, in a somewhat elliptical shape. He was positive some creature had made it, owing to the consistency of its shape and finish, but what kind of creature that might be was still a mystery. He had managed to engage the attention of an archaeologist who had also been puzzled, but it was agreed that running water could not have created the tunnel in the sand. Where it was still intact, they discovered that it took a turn of more than 90 degrees, widening to four feet at the bend, like a den or turn-around area.

When I quizzed Ed again about the lining of the burrow in the summer of 2009, he reiterated his recollection of a hard surface that constituted the interior of the tunnel, "as if some animal had brushed against it again and again." The thickness of this lining that he broke up with his loader for over 40 feet he estimated to be between one and two inches, enough to provide the needed strength to keep the sandy voids from collapsing.

As for fur keeping the interior smooth and shiny, I am reminded of the description a farmer gave me of the creature he saw on the ice of a Saskatchewan lake. He thought it was just a shaggy, poorly-tied round hay bale, so from his description I visualize these creatures polishing the liner with their long fur on each pass, perhaps leaving a film that in some way helps preserve the integrity of the lining.

I had heard of another unusual hole in the ground the previous winter, and was able to discuss it on the same trip with someone who had seen it. So, again I found myself sitting at a kitchen table, and again the tunnel's interior was compared to the smooth table top. A few years earlier, this hole had suddenly appeared over winter in a farmer's grain field near Amaranth on the western shore of Lake Manitoba.

When the farmhand, whom I spoke with later, had worked the land in spring, he discovered a four-foot diameter, almost vertical shaft that narrowed down to about three feet where it began to slope in the direction of the lake. The two-foot layer of overburden that had collapsed was nowhere to be seen, having slid out of sight on the smooth surface.

Many locals had gone to see the tunnel before the farmer filled it in with stones and dirt, and I found that everyone's description of it was the same. One individual said the shape of it reminded him of the funnel of a tornado. No one had any idea what could have made it, but some of the witnesses of Native background would mention big snakes.

Strangely enough, I heard the same explanation over and over again on the Sioux Valley Reservation. Elders everywhere seemed to be of the opinion that huge snakes inhabited the tunnels. That such snakes existed I had no doubt, as I had heard of dozens of sightings around the province. That large snakes might live in the tunnels was a reasonable possibility as well. However, that the snakes might be capable of creating the tunnels, I had serious doubt.

So, the mystery of what made the tunnels plagued me for some time until the answer came that first winter when I traveled throughout the north collecting stories of unusual creatures. The giant beavers that the folks at South Indian Lake spoke of, and the sandy tunnels that they were known to live in, immediately answered the question of who made the tunnels in southern Manitoba.

I spent time on three different occasions trying to dig into the collapsed tunnel near Swan Lake, until finally it was agreed that the burrow must have ended where the last section had caved in. So, much as I wanted to see the interior of one of these burrows, I was not able to do so.

It was in the spring of 2005 that I got the break I was hoping for, except that I didn't realize its significance at the time. I was paying my first visit to the Birdtail Sioux First Nation in the western part of southern Manitoba. My typical enquiries about unusual creatures led me quickly to a young man who, the day before, had had a very close and dramatic encounter with a Sasquatch. But, it was a comment from a man who had grown up at Sioux Valley that soon altered the course of my research. He recommended that I visit his mother, and get her story about the tunnel that a backhoe exposed in her back yard when it dug a little garbage pit for her some years earlier.

Unfortunately, by the time I made it to Sioux Valley that fall, she had passed away. Neighbors referred me to another location where a similar tunnel had been exposed, so I spoke with a number of people who remembered seeing it while it was exposed. Apparently, residents were so intimidated by the large cavity that they

A caved-in shaft that appeared in a farmer's field near Amaranth. Note the protective cribbing plastered throughout, with its broken edge visible on top.

asked for it to be filled back in. Again, snakes were the creatures that were thought to inhabit the tunnels.

I realized that I was almost certainly in an area where giant beavers *had* lived. It took me a while to accept the possibility that they perhaps still lived there, even after one elder described the big eyes and whiskers his grandfather spoke of.

So, by April of 2006, I had come to the realization that if the critters had been seen a generation or two ago, and their tunnels were still intact, then perhaps *Castoroides ohioensis* still lived here. That is why I, with only faint hope, decided to camp beside the Assiniboine River near where someone else had camped the summer before. The sun was down, and I would soon retreat to my bed in the Jeep. I stood motionless, still gazing upon the river, the darkness not yet diminishing my view, wishing to behold the creature that I believed lived there! However, such hope was not a realistic expectation.

But then something came into view directly in front of me, and my eyes fastened on something that matched my dreams. I gawked in disbelief! A large form appeared clearly on the surface of the water, even though small twigs and branches came between. Instinctively I bent down to gain a better perspective, and as I did so, the large head went down and the distinctive slap of a beaver's tail followed.

Large deep hole in a creekbed in northern Manitoba believed to be evidence of the giant beaver.

I was shocked. Had I really seen it? This was almost too surreal to be true. I pondered the incident, not only all evening, but also for weeks and months, asking myself if I really saw a giant beaver.

What *did* I see? The part that stands out in my mind is the big head, at least the size of a basketball. Behind the head, the animal's back was also above the water, and behind that, something else—presumably the tail. One long glance was all the time I had—not enough to pick out any details. The overall length was an incredible seven to eight feet. It seemed to float with the current, which moved faster than a walk. The creature must have seen me standing there, and in fact probably had been spying on me all along, and surfaced to get a better look. And in those brief moments, I also got a look—and will be forever grateful.

If I had not specifically gone looking for the giant beaver in this particular spot because of the stories I had heard in this community, I would be doubting my eyes and ears to this very day. But since I saw what I went there to see, I am fully convinced that I had an encounter with a very much alive giant beaver. And I am not by any means alone in this experience.

I have spent many hours beside that river since then, before dark and also at first light, hoping to get another glimpse, to no avail.

I have come to the conclusion that giant beavers must not surface often, at least not in the daylight. When I finally got to meet the man who had camped beside the river the previous summer, he insisted that he had not seen anything unusual. I was extremely disappointed. When one last time I asked, "Are you sure that you never saw *anything* out of the ordinary in the river?" he conceded that there was this one time when something "like a little submarine" swam just below the surface, pushing water and making waves.

That admission made me feel a little better, but it reinforced the notion that there was good reason why these creatures had not been seen by many people.

Speaking of making waves, a similar phenomenon comes to mind. I have already made reference to someone who lives beside a lake in western Saskatchewan. The farmhouse is set on the high bank overlooking the lake, offering an excellent view. The farmer reported seeing a large creature swimming around, leaving a wake much like one that a small motorboat would make. He also mentioned seeing an unusual object on the ice one spring. It resembled, he said, a small, round, bushy hay bale, but he knew it was not one of his. He wondered how it could have gotten there, and decided to go check it out. However, before he was able to do that, it had disappeared. He noted that there was a thin strip of open water along the shore.

His observations convinced me that the giant beaver lived in his lake, and perhaps in many other lakes and rivers across the continent where high sandy banks afforded the appropriate soil conditions for their burrows.

Another interesting phenomenon that was observed in that lake from the air may provide additional clues. Years ago, a contractor flying over the lake noticed what appeared to be holes on the bottom. Efforts apparently had been made to locate them from a boat, but no further information has come forward to date. However, this story prompted me to modify my theory somewhat in regard to the burrows. I had previously assumed that the giant beaver burrows began underwater on the side of a bank, similar to those of common bank beavers. But, now I entertained the idea that *Castoroides ohioensis* perhaps burrowed straight down from the bottom of lakes and rivers before angling laterally. A few comments I noted from different individuals seemed to corroborate this view.

Some told of shoving long sticks into underwater holes. Another related an incident in Lake Manitoba where a number of men had been swimming, seeing who could dive the deepest. One frightened man surfaced after his partners had become worried about him, and suggested they leave the area. He described entering a hole whose sides he could feel as he worked his way out.

In the summer of 2006, after I glimpsed what I believe was the giant beaver, I decided to investigate the tunnels that were familiar to the folks at Sioux Valley.

The man who now lived in one of the yards where the digging of a garbage pit had exposed a burrow had no objections to my drilling on his property, so I began to explore different methods of locating the tunnels that were remembered to be about eight to ten feet below the surface.

I was convinced that there must be some technology that could identify voids, so that line of enquiry led to an introduction to ground penetrating radar, or G.P.R., a very expensive system that, in suitable soils, should be able to show the difference between solid soil and voids. During my search for a government department that might have such an instrument, I was offered the use of a small hand auger, which, with extensions, could reach to a depth of ten feet. So, while I waited for a G.P.R. to become available, I began digging.

Stewart (the owner of the property) and I started where the majority of witnesses remembered seeing the burrow at the bottom of the garbage pit. We chose a spot well away from the damaged area, or so we thought. When we reached our maximum depth and found no void, we moved aside two feet and tried again. Then two feet on the other side, and back and forth again. Curious passers-by would drop in. Some who remembered seeing the tunnel would recall its being over further in one direction, so we would concentrate there until someone else would come along and insist the tunnel was in the opposite direction. One lady remembered it being on the opposite side of the house, so we dug ten fruitless holes there.

An older gentleman, Albert, who was one of the respected elders of the community, dropped in now and then, and gave us a good deal of encouragement. It was his son who, as a boy, had been sent into that tunnel in the uphill direction, with a rope around his waist. In the downhill direction, towards the river, it was well known that a small dog had ventured, never to be seen again. Its yelps and cries were still vividly remembered.

Albert was very interested in our efforts, and enjoyed discussing the project. He said he used to live further downhill from where we were digging, across the road beside the river flat, a little over a stone's throw away. He had been anxious to move away from there, and explained why.

He decided one day to take the lid off his new well in order to look inside, since he had been away working when the waterworks were installed in his home. After he unbolted the cover from the 30-inch cribbing, he was shocked to hear and see something at the bottom of the well "like a war going on down there." He closed up the well, and made plans to move.

The well driller told him that his machine had encountered a void somewhere below ground, and when I called him, he said that it was virtually unheard of to drill into an open space underground like he did there at Sioux Valley.

STRANGE CREATURES SELDOM SEEN

Since that well was directly in line with the previously exposed tunnel, we moved to that spot, and dug another ten holes. Then we moved back to where we started, and dug some more. My arthritic shoulders had complained right from the outset, so younger muscles were employed to help with the task. Some days three young men took turns manning the auger. The holes we dug were two-and-one-quarter inches in diameter by ten feet deep, and each hole took a good half hour to dig. So, by the time we had dug our sixty-ninth hole at the end of November, my enthusiasm as well as my pocket book were greatly diminished, and I hoped never to use the hand auger again, except for single holes into predetermined voids.

On that last afternoon, the day before winter weather set in, I had been joined by Albert's son, Fraser, who had been inside the tunnel as a boy. As the other men were digging the last of the holes further downhill, Fraser and I discussed our options. When he learned that I had brought with me a 15-foot-long by three-eighths-inch-thick sharpened rod, he suggested we use it in the holes that had been dug. So, since he had the energy and enthusiasm, I agreed to it.

He thrust the rod down into the first of the 41 holes that stretched across the yard. It stopped abruptly in the firm sand. The same happened in the next few holes—until we reached the spot where the majority of witnesses believed the tunnel to be. There the rod went down easily in soft sand, over three feet past the level where the auger had stopped. Finally, a new development—at the very last hour!

We reflected on what that could mean. My boyhood experience with pocket gophers came to mind immediately. Those creatures also dug burrows, barely bigger than the shafts we were digging with the auger.

The dirt piles they kicked up were a real nuisance on the farm, especially in the garden. For five or ten cents apiece, I had been encouraged to trap as many as I could, and there began my only successful trapline. I soon discovered that pocket gophers, or moles as we called them, would occasionally notice the trap I set underground just below the mound of dirt. Instead of getting caught, they would plug the tunnel with dirt, trap and all. Instinctively, my thoughts compared the actions of the pocket gopher to what we had encountered—a pocket of soft sand inside the alleged tunnel. I could not help but believe that the giant beaver had plugged up its burrow for some distance where the damage had been done to it years ago by the backhoe. It seemed perfectly reasonable that a creature would maintain the integrity of its lifeline in such a way, and I felt a measure of satisfaction in the discovery. We had not found the void we were looking for, but we found some evidence of it.

In our excitement, however, what we neglected to do was test the holes that had just been dug that day, holes that lined up with the one full of soft sand. Therefore, something significant remained undiscovered until the following summer.

Since there was some hope that I might gain the use of a GPR instrument for a day in the summer of 2007, I confidently returned the borrowed auger.

I discussed the effectiveness of GPR with the supplier of the instruments in Toronto, and, upon hearing that we were continually digging through about a foot of clay in each shaft, there was concern that radar might not penetrate this layer well enough to outline the tunnel below it. A soil test to determine the density of the clay was recommended, so over winter this was accomplished. The supplier, after considering the results of the test, did not believe that GPR would be reliable in the circumstances, but only a trial would tell for sure. So, hand drilling once again became the operation of choice, with some hope that a GPR instrument might become available and might just work.

If it did, then we could determine exactly where the voids were, drill down, and photograph the interior of the tunnel. A friend loaned me his infrared, underwater camera that just fit into the 2 ¼ inch shafts we were drilling, so we had the ideal device to view and photograph what we might discover.

Drilling for giant beaver burrows at Sioux Valley.

By this time I had also been promised the use of professional equipment that was used by a Winnipeg company, Uni-Jet, to photograph the interior of water or sewer pipes. Cameras mounted on various sizes of self-propelled wheels controlled remotely from inside a big truck could yield clear video pictures more than 1,000 feet away.

Prospects for finding a tunnel and exploring it in 2007 seemed good. Meanwhile, I had made some excursions into the north where a few accounts of big beaver lodges had surfaced. There was one spot I especially hoped to see. A trapper traveling around his family's remote, traditional trapping area north of South Indian Lake had seen a giant beaver there a few years earlier. A bigger than average beaver lodge was nearby, with full-sized trees on top of it.

Access to it, I found, would be too strenuous and dangerous by boat and canoe, and the cost of a helicopter was prohibitive. The little lake where the lodge was seen was also too small to chance a landing by small aircraft, so I opted to take a shorter three-day trip by boat and canoe to another area where a giant beaver had recently been seen.

It was a most interesting trip into the pristine wilderness, but a giant beaver house was not to be found. However, my guide pointed out that there was a sandy ridge nearby, so possibly its residence was instead the typical burrow about which the folks from South Indian Lake had told me.

Since summer was not the ideal season to reach that other remote, but favoured lodge, we made plans to travel there by snowmobile the following March, when the days would be lengthening and getting milder. Unfortunately, when the time came, a storm dropped four additional feet of snow on the area, so that clinched the prospects of visiting the site in 2007.

I did, however, attempt to check out several less remote sites in summer, where giant beavers or their lodges had been seen, but none of the trips were successful as far as concrete evidence was concerned. However, all of them were rich in new stories about the creatures.

One account came from a family I had been referred to as having seen a giant beaver. They lived on the high bank of Footprint Lake which surrounds the community of Nelson House. In the fall of 2006, the family watched in awe as an unusual creature appeared in the water below them for about half an hour. Its 13- to 15-inch-diameter head would be visible for five to ten minutes at a time as it swam "looking around," and then it would disappear. That is when fish would begin jumping out of the water, "like reverse rain," dozens at a time. Then its head would reappear for a while, and when it dove down again, the fish would begin jumping again. Never before had they seen or heard of such a thing.

The question for me was, do fish jump out of the water only if they are being pursued? If so, would the giant beaver (assuming that is what it was) chase fish if it had no intention of catching them?

A neighbor also mentioned seeing something which made a large wake, swimming below the surface.

There are numerous other stories of unusual sightings at Nelson House. Often they are reported by mothers who are watching their children swim along the base of the high banks. The terror of the moment causes immediate evacuation, lasting only until the children's desire to go swimming outweighs the fear some days later.

I firmly believe that one of the creatures seen around Nelson House is the giant beaver. Many witnesses describe a good-sized creature that they see swimming in the daylight or in the dark—a creature with a big head, and either wearing big whiskers or carrying a bundle of sticks in its mouth. A large hole was visible for a while one summer, high up on an eroding bank, lending further evidence to the giant beaver's presence.

I do believe, however, that there are other creatures in Footprint Lake as well, based on descriptions from a variety of folks who have seen them. You will find them mentioned in other chapters.

Returning from one fruitless canoe trip near Nelson House, we stopped to chat with a retired trapper at his cabin in the woods. My guide and his wife asked him, in Cree, if he had ever seen a giant beaver, and he described to them what he once saw. Then he went on to tell of a trapper he had known who killed a giant beaver. Not wanting to transport the bulky hide all the way home, he had cut it up into five blanket-beaver-sized pelts. As we left that place, my guide mentioned that his grandfather had once done the same thing, except that he had cut his hide into four pieces. I heard later that he had actually caught two huge beavers on the same trapping trip, yielding eight ordinary-sized pelts in total.

As I motored further north early the next morning, a creature I didn't immediately identify ran in front of me along a causeway. I knew it was too big to be a fisher, but its dog-like gallop made me think of an animal that the Native people in a variety of communities had described to me. A sea-dog is how they translated the name, and these creatures were known to come out of the water on occasion and run around like dogs. (See the chapter on Sea Dogs.)

This animal disappeared into a treed area, from where it emerged right after I stopped and readied my camera. I was able to take some pictures of it from close-up, and it was then that I realized I was face to face with a much-feared wolverine. Fortunately for me, it hesitated only briefly before resuming its puppy-run back the way it had come.

STRANGE CREATURES SELDOM SEEN 33

Wolverine on and beside the road in northern Manitoba.

The experience reminded me of two other wolverine encounters that also could have proved dangerous, and they both took place right near where I lived, less than half a mile apart. The first took place on the reservation, across the river from my campground. A student of mine related how she had walked into a small building in her grandparents' yard and had surprised a wolverine which had probably been attracted by the smell of meat. Hazel had jumped on the bed, screaming for help. Her grandmother, who had been working in the garden nearby, came running, and the creature fortunately ran away.

About ten years later, at the back of my campground, a young farmer was cutting alfalfa with a tractor and mower. A strange animal had run out from under the tractor, when it paused and seemed to prepare to attack. Sensing the danger, Derek grabbed a hammer and watched as the creature took a run at the tractor, jumped, and almost cleared the top of the back wheel. Fortunately, it fell back from the moving tire, and ran off.

Later that morning I picked up a man, Tom Bird, who was looking for a ride into Leaf Rapids. He lived alone in the wilderness on the shore of Suwannee Lake, and once a month he would catch a ride into town to pick up his cheque and supplies. He told me of a huge beaver house he had seen, and promised to take me there in his boat.

So, the next afternoon, loaded with supplies for himself and his dogs, we returned to where he kept his boat, and set off for his camp. After unloading his supplies and combing the shoreline a good hour for his too-well-hidden cache of gasoline, we started off in search of the giant beaver house. It was choppy where we had traveled, but nothing like we encountered around a point where the northwest wind pounded our boat mercilessly. Realizing that discretion was the better part of valor, we soon turned back with the intention of trying again in calmer weather.

When I returned about a month later, I discovered that it was a good thing that we had not been able to make the earlier trip, as the distance to the site had been considerably misjudged. I met Leslie Baker, the man who had been the original guide to that beaver house in 1987, and it was his story that made this seemingly futile trip worthwhile.

Leslie said that his father had known the trapper who told of a huge beaver lodge in his trapping area, far from the settlements where the men would live between trapping seasons. Many refused to believe him, having never seen such a thing themselves. Leslie's father never made it to the site, but he remembered the story well enough to describe the location to his son.

So it was that, in 1987, Leslie and his older friend Tom Bird went to see the lodge. It had obviously been abandoned long ago, as the huge roof was caved in. Leslie estimated the diameter of the interior to be about 25 feet, and the overall length between 40 and 50 feet. Logs in the feed pile were over a foot in diameter, and six to eight feet long. A fire went through the area a year later, so it is not likely that any evidence of this extraordinary lodge remains there to this day, nor that of the 20-foot-high dam seen about 60 years ago at McKerracher Lake, that was made of long 18-inch-diameter trees, "like hydro poles" stuck into the mud.

When I returned later that summer to the place of the 69 holes at Sioux Valley, I took with me the long rod that had helped discover the one soft shaft the previous fall. I decided to put it down into the last few holes that we had neglected to test, just in case they held some surprises. To my astonishment, the very last one did. My 15-foot rod fell down 13 ½ feet into a void where we had dug down only ten feet six months ago. I tapped the rod up and down to feel the bottom, and it was like I was touching a concrete floor. I calculated the depth, subtracted the ten feet we had drilled, and realized that the tunnel must have been just below the ten foot mark where the auger had stopped.

I was delighted. Now I would be able to get a picture of the tunnel with my friend's little infrared underwater camera. Those pictures should finally be able to prove the recent, if not present, existence of the giant beaver. Besides, they would show that the creatures lived in tunnels—something that none of the limited scientific information ever mentioned.

I should have brought a witness to the scene. In fact, I should have raced home the four hours to borrow the camera, and another four to return. But instead, I decided to bring the camera on my next trip in a month's time, when I had another meeting in the area. I simply did not realize the dynamics of what I had just experienced. That is why, when I put my rod down the hole a month later, I was devastated to find that it went down only half way—an unbelievable half way! The sand was soft, but it filled the void that had been there a month ago. Not only that, but the sand reached up into the shaft a few feet as well!

Skeptics kept suggesting that the sandy void had just caved in. But, if that had been the case, then there should have been a huge cavity left above it, perhaps right up to the surface of the ground. Such was not the case, and there was still the question of the two-and-one-quarter-inch shaft being full of sand several feet above our original depth.

I felt that I knew the answer, based again on my pocket gopher experiences. I was convinced that the giant beaver, finding that we had drilled into his tunnel, didn't appreciate the exposure to the open air (even if it were ten feet above), and filled in the whole area with sand. One question, however, remained. Why was the shaft open to the tunnel one day, and a month later, blocked solid? Why hadn't the critter filled it in over winter?

A good deal of pondering took place before I was convinced of a plausible explanation. Assuming that *Castoroides ohioensis* are not procrastinators like *Homo sapiens*, they would have repaired the damage upon discovery. Therefore, the damage could not have occurred until recently, and that would have taken place when I put the rod down a month earlier.

I do not recall feeling any obstruction when I lowered or dropped the rod into the shaft, but there may have been a thin layer of sand that the rod dislodged, exposing the tunnel to the air. Why else did the creature wait until I had been there before it filled in the void? Perhaps we will never know just what happened, but the experience taught me one thing, and that was to have my cameras on hand, ready to use before something came along and messed things up.

Disappointed as I was, I hired men again and resumed drilling, 15 feet downhill from that remarkable hole. I considered having a mini-well drilling rig built in order to make the holes faster and easier, but several factors discouraged that plan. For one thing, it would take time, and cost a good deal. Besides, I told myself that once we found a void, which could be any day, I wouldn't need the contraption any more. And, GPR might yet work to locate the voids.

So, we dug on . . . and on. We found nothing by drilling down 10 feet, so, since extra extensions had been made available, we went down 13. And then 16. And just in case the tunnel took a dive, we drilled down 19 feet. Nothing!

We packed it in, agreeing that if we were ever to try again, it should be uphill from where the backhoe had exposed the tunnel.

Another meeting took me to the area about a month later, and that encouraged me into trying once more. Surely, by drilling over 100 feet away from the original damage, the burrow should be open and accessible. I brought my friend's camera, just in case, to take some quick pictures of the tunnel by way of the two-and-one-quarter-inch shafts. I also brought my own four-directional underwater camera which would be able to shoot both up and down the tunnel if we found it. To accommodate the bigger camera, we would need at least a six-inch diameter hole, so I purchased a hand auger that size and had extensions made for the handle.

This time I was going to be ready. If we broke through, I would have cameras in place to capture the action if and when the creature would come along to plug the intrusive hole. I realized that the cameras would be destroyed in that process, but I counted on having enough footage recorded before that happened to prove the existence of the giant beaver. I had arranged for backup assistance, as I planned to man the cameras until there was some action, regardless how many days and nights it might take. We didn't lose any sleep at all.

Fifteen holes later, each one dug to a depth of 13 feet, déjà vu!

Three positive developments came out of the effort, however. The first was that we were recording our activity. I had long felt that it might be of value to create my own documentaries, so I had engaged a young cameraman to be on site for a day. (In fact, he happened to be at the dig site just when the auger fell almost a foot into a void, and then into more sand. I hoped that this was a sign that we had accessed the edge of a burrow, and that the next hole, two feet over, would hit the centre of it, but such was not the case.)

The second positive outcome came after everyone else had gone home. I was finished for the year except for the test with the rod. And, like the fall before, I struck soft sand. Of the 15 holes, four were soft. They were side by side, except for one solid hole separating the two pairs. Since we had dug them less than two feet apart, it was probable that the first two holes accessed one tunnel, and the next two accessed another, a foot or two apart. The virgin soil in the hole separating them seemed to indicate that there was perhaps a Y in the tunnel, or some other configuration yet to be considered.

The third encouraging factor was the absence of clay. Whereas there had always been a substantial layer of dense clay where we had dug further down hill, here there was no clay to speak of, not even to the 16 foot depth. If the GPR were to work anywhere, it should work here where only sand existed.

Whenever I would mention that the tunnels were made in sandy soil, alert thinkers would express their concern, and rightly so. Sand is definitely not a stable material, so the common impulse is to dismiss the whole idea as a ridiculous myth.

But this is where we mortals so easily err. We tend to think that everything needs to make sense from our limited perspective, and therefore we ignore the facts. And one fact seems to be that the interior of the tunnels are smooth, not rough and sandy. Every eyewitness I had heard from spoke of a hard, smooth surface.

One lady, whose son had climbed into the tunnel near Amaranth, had taken a picture of it, and the photo reveals what appears to be a liner about an inch thick resembling clay. Certainly these creatures have access to whatever soils that exist near a waterway, so it is entirely possible that one of their God-given instincts is to keep a weak structure from collapsing by use of plaster. *Castor canadensis* certainly does a remarkable job of making its dams and lodges secure with the use of mud, so it is definitely within the realm of possibility that *Castoroides ohioensis* was endowed with enough sense to make its abode secure as well.

It will be interesting to discover, some day, where the creature stores the sand that it excavates. Does it use its huge teeth or its paws to do the digging? Or both? Does it carry a huge mouthful of sand back to the river, and return with a mouthful of construction-grade clay? (And perhaps a stomach-full as well?) How many beavers would participate in this exercise? Do they enlarge the tunnel every so far to act as passing zones? What about oxygen? There must be an adequate amount of it underground, or else they would have appreciated the little opening we made for them.

As for swimming long distances underwater—they must be capable of staying below the surface for a long while, since the river, where we are operating, is about half a mile from where the tunnel rises above the level of the water. It must be that they swim through a water-filled tunnel for that distance, holding their breath, with their mouth full. If they carried the sandy material other than on the inside, it would quickly wash away.

One day, in the summer of 2007, I decided to check the river bottom for any indication of large holes, so I used a depth-finder attached to a small boat to survey the area where I suspected that the tunnels might begin. Sure enough, there were several spots where unusual images appeared on the monitor that were quite different from the normal river bed.

Interpretations of these images which I photographed did not support my expectations of holes in the river bottom, but more likely represented images of mounds on the river floor. I came to the conclusion that perhaps the amount of sand the beavers release into the river would create a noticeable mound near their tunnel entrance, and would therefore show up on a depth-finder as such.

In the summer of 2008, I again put the depth-finder to work on the river bottom, but, lacking expertise and time, no definitive discoveries were made. Perhaps a view from the air might have shown the presence of large holes similar to the one we saw in the far north from the helicopter, but the silty water would probably not be clear enough.

The most puzzling question for me, as I have pondered the existence of the giant beavers in the Assiniboine River, is in regards to their food supply. Speculation by scientists in the limited writings about the giant beaver assumes that they fed on vegetation in swamps and ponds.

However, unless I am mistaken, which is a definite possibility, I can not think of anything for them to eat in that river other than fish. But beaver and fish just don't sound compatible, somehow. However, the fish-jumping account out of Nelson House puts the question into a whole new perspective.

They definitely don't cut down trees in the area, and the few that grow along the riverbank, would their roots be plentiful and deep enough to sustain a population of these large creatures? And would the trees be able to survive?

Surely there would not be adequate nutrition in the soil itself?

Vegetation in the river seems virtually non-existent, although not enough time has been spent investigating this. I initially assumed that there had to be something growing underwater or underground for them to feed on, but after I tested the bottom with a huge fishhook and found nothing, that theory was quashed.

The water in the Assiniboine River fluctuates a good deal from season to season. In spring it is typically high, and by fall, unless heavy rains occurred on its watershed, it is low. There is virtually no vegetation visible along its muddy banks, and no indicators that there is any creature living there other than the ordinary beavers and some raccoons.

I spent a whole day in 2006 examining miles of the winding river banks from a boat, and there was no sign of the presence of *Castoroides ohioensis*. The river at that time was at its normal summer low stage, and I hoped to spot a hole in the bank, or large tracks in the mud, but nothing was noticeable. I was not that disappointed, for I knew that if these signs were ever present, then other people would have been aware of them too.

Some stories indicate that the big beavers do come ashore on occasion, however. In the north, where numerous sightings have taken place, it appears to be more common, but along the Assiniboine and other southern waters, it must be a very rare occurrence.

One young man who was helping me with the digging related an experience he and some of his friends had a few years ago. They had been approaching the Assiniboine River at their favorite fishing spot at the junction of the Oak River when they heard a loud "plop." He compared the sound of it to a large drum of fuel being dropped into the water. They quickly ran to see what had made the noise, and saw something big swimming just below the surface, creating substantial waves. When the creature had reached the Assiniboine a short distance away, it had disappeared into the deeper water. Another man from the community remembered seeing something with a big head swimming in the river, carrying what appeared to be a large log in its mouth.

From the boys' experience, it appears that the beaver must have jumped into the river from on top of the bank when it heard them approach. The tendency of this creature to jump when surprised reminds me of the experience Wilson Moore of Nelson House related to me. When he and some others had been out hunting in a boat one spring, a large black animal they thought was a bear came out of the woods towards shore. One member of the party was interested in taking the animal home, so Wilson shot it with his big-game rifle. He told me that when the bullet struck, the animal had made a huge bound forward, collapsing at the edge of the water. It had been so large that two of them had not been able to lift it completely off the ground.

The account of the experience the boys had beside the Assiniboine River corresponds with the one given by the gentleman who had camped a whole summer beside this river, very near where the boys had their experience. He too had seen only the "submarine" creating waves.

Similar reports are coming to light from different communities, as this one from Nelson House portrays. A boat-load of hunters was returning from an excursion up a shallow river when they noticed something large swimming ahead of them for a long distance just below the surface, making substantial waves. When they entered the deeper waters of the lake, the creature disappeared, leaving them all puzzled.

Legends of the giant beaver span a good portion of the North American continent, especially among certain Native tribes. The Nisku of Labrador, for example, have accounts of the beaver from the recent past, and even have a river they named after it, which translated would be "Giant Beaver River."

Nova Scotia has ancient stories of them as well, as do other maritime provinces, Canada's Territories in the North West, and Alaska. A hurried trip along the Alaska Highway a few years ago netted me three giant beaver stories with little effort on my part. I believe that in any area of the U.S.A. and Canada where these stories persist, there is a good possibility that the creatures themselves may still live there.

Two hundred years ago, in 1808, Alexander Henry the Younger was traveling northward down the Red River in present day Manitoba, Canada, on a trading expedition and he tells of meeting a Native man near the forks of the Red and Assiniboine Rivers where the city of Winnipeg now stands. The following paragraph from Henry's journal focuses on a giant beaver that was seen in the vicinity:

> "A Saulteaux, who I found here tented with the Courtes Oreilles, came to me this evening in a very ceremonious manner, and after having lighted and smoked his pipe informed me of his having been up a small river, a few days ago, upon a hunting excursion, when one evening while upon the water in his Canoe, watching the Beaver to shoot them, he was suddenly surprised by the appearance of a very large Animal in the water. At first he took it for a Moose Deer, and was preparing to fire at it accordingly. But on its approach towards him he perceived it to be one of the Kitche Amicks or Large Beavers. He dare not fire but allowed it to pass on quite near his Canoe without molesting it. I had already heard many stories concerning this large Beaver among the Saulteaux, but I cannot put any faith in them. Fear, I presume, magnifies an ordinary size Beaver into one of those monsters, or probably a Moose Deer or a Bear in the dark may be taken for one of them as they are seen only at night, and I am told they are very scarce."

Henry is typical of so many whose only motto is "seeing is believing." If I had stubbornly maintained that philosophy, I would not have been able to gather stories like I have about the dozens of creatures that people claim to have seen. Whether or not that would have been a good thing is debatable, but to be honest, I have been delighted to enjoy the privilege of hearing so many accounts about unique creatures. In fact, I share the excitement of that Saulteaux hunter, and am bursting to share my knowledge for the world to enjoy. Unfortunately, like that hunter, I find disappointing skepticism at every turn.

Wouldn't it be exciting, though, to find that these creatures still live in Manitoba's capital city! It appears, however, that I may have to drag a specimen into the open before the giant beaver will be accepted as extant. Perhaps then there can finally be dialogue and concerted research into the ways of this and other elusive creatures which only a few people have come to know about thus far. It has been almost a decade since I first realized that the giant beaver was not extinct after all, and yet not much has changed in that period of time in regards to its acceptance as a living species.

I am surprised that out of the centuries during which the fur trade was carried on there wasn't more documented evidence of this oversized beaver. Besides cutting up the occasional giant beaver hide into smaller-sized pelts, perhaps some of the unusually large furs were saved for use as blankets, or for making articles of clothing.

It was brought to my attention that when registered traplines were established near the middle of the last century in Manitoba, there was a period when trappers were asked to document every beaver house in their territory. The beaver population had declined seriously, and therefore measures were taken to prevent the species from dying out. With the scrutiny afforded to these creatures and their habitat, one would think that the presence of the odd giant beaver lodge in the north would have come to light. Obviously, this did not happen. The reason, I feel, is that there were so few of them, scattered in such isolated areas, that they were hardly noticed.

I believe also, from reports I collected, that trappers did catch some juvenile giant beavers, thinking them to be ordinary ones, even though they were darker in color. Furthermore, when traps and snares were destroyed too often in certain lodges, they were just abandoned by trappers as a lost cause. The prevailing assumption of those bygone days, much as today, was probably that these larger-than-average beavers were just anomalies, and not an altogether unique species. There were too few of them caught for them to be accepted as legitimate animals of North America.

The end of December 2007 revealed a giant beaver story that was as surprising as it was refreshing. It pointed to an area that I had never expected to be big beaver habitat, even though I had observed the potential and thought on it briefly as I passed by it through the years.

When traveling on Highway 60 between Grand Rapids and The Pas in central Manitoba, the road follows a sandy, esker-like ridge along the north shore of Lake Winnipegosis. Where the lake can be seen, it is obvious that it is far below at the base of the isthmus which separates Cedar Lake from Lake Winnipegosis. The high sandy bank appears to be ideal for burrows, but I had never heard a word about a creature in that lake that made me suspect their presence. Not until I visited the Indian Birch First Nation earlier that fall.

I had gone there to obtain a drawing of a horned serpent which had been seen in Swan Lake (a different Swan Lake!) some years ago. Since the witness was not at home, I went to see an elder whom I had met on a previous visit—baseball Hall of Famer, Buddy Brass. A friend of his dropped in while I was there, and entered into the conversation about unusual creatures. He mentioned two things that were significant to me. The first was that he had seen where large poplar trees had been gnawed off about six feet above the ground by something with very large teeth. This was beside a river that drained into Lake Winnipegosis. I did not get too excited about this information as I had heard and investigated similar information involving beaver stumps cut six feet above ground only to find that a local flood had raised the water level, allowing the ordinary beavers to cut down trees from on the ice far above the normal ground level. The large teeth marks made this story intriguing, however, and plans were laid to visit the site by boat in the future.

The second bit of information he gave me got me more excited, though. He told me that his wife, who lived at Easterville along Highway 60, had seen a big beaver as it was crossing that highway.

Just before Christmas, I had a chance to drop in on her, and get her story.

Three or four summers before, Mary and her daughters were motoring along Highway 60 about half an hour west of Easterville. During a pit stop, they had seen a large beaver amble across the highway from the north to the south in the direction of Lake Winnipegosis, a few hundred yards below them. It had taken up half the highway, dragging its typical beaver tail behind. Her hand gestures also indicated an average giant beaver-sized creature, and when I asked how it compared to a bear, her eyes lit up and she exclaimed that it was a good comparison that she had not thought of before. There was no doubt in her mind that it was a beaver she and her three daughters had seen, and she had told them forcefully that they were not to mention the experience to anyone, as they would not be believed.

As we continued our hours-long conversation, she asked if her husband had mentioned his unique experience involving some huge birds. Since he was home, he came and related his unusual story, which is chronicled in the chapter on Big Birds.

His wife's remarkable giant beaver sighting related to me in December of 2007 answered my question regarding such observations on or beside highways. I wondered why I had never heard of any before, but since this one had been on a lonely stretch of road in the middle of a vast wilderness, I was not that surprised. Within a few months of hearing her story, however, I was surprised on three more occasions.

In the summer of 2008 a young man from Nelson House who had heard of my interests tracked me down in Thompson in order to share his story. He and a carload of young people had been motoring to that northern Manitoba city some years ago when he spotted a large beaver up ahead on the side of the road. He had stopped, grabbed a baseball bat, and commented that he was going to get supper. As he approached the creature, he realized that it was much larger than an ordinary beaver, and about that time it had gone on the offensive and started after him. He had barely managed to escape into the car, and the other occupants had told him that it had taken huge jumps behind him.

A second sighting took place in the city of Brandon, of all places. But, why not, since the Assiniboine River, which I believe contains the creatures upriver at Sioux Valley, runs through this southern-Manitoba city. A young lady who had lived in Brandon for a few years told me she had been returning home late one summer night and used a road that took her through a park near the river. What she at first thought was a big garbage bag on the side of the road turned out to be something alive, and when it crossed in front of her car, she saw in her headlamps something that reminded her at first of a bear because of its size. In her own words, Jena Grandmont described her experience:

"It walked very slowly . . . there was no urgency for him to get off the road and out of my path. It was kind of a saunter back and forth. It was dragging on the ground. It was hard to distinguish the object's shape. It looked black and very shiny/wet but it was dark out. I remember thinking, what the heck is that! It was not a recognizable animal to me, and it wasn't rushing to get out of the way, so I rolled down my window to see it better and then when it started moving and I realized I had no idea what this thing was, I quickly rolled my window back up because I was honestly a little frightened. I thought at the time that the closest animal it resembled was a small bear so I thought I better leave it be in case there was a mother bear around. But when it made it off the road and over to the grass (and out of my headlights) I could then see what looked like a beaver's tail dragging behind it . . . so I assumed it was just a *huge* beaver. I definitely told people about the encounter over the next few days that passed, so it was on my mind as an unusual experience, but I have never spotted anything else remotely like it and I am always on the lookout for wildlife."

The animal's shiny, wet appearance suggests that it had just come out of the river. Being the Assiniboine River, downstream from where there is ample giant beaver evidence at Sioux Valley, (besides my own sighting there), this story seems to indicate that the creatures occasionally visit the big city, if, in fact, they don't actually live beneath it in their deep burrows.

The third sighting, and the one that surprised me the most, was made a few years ago practically in my back yard. I live near the shore of Lake Manitoba where there are no banks to speak of, as the land slopes very gently to meet the lake. Not in my wildest dreams did I expect this to be giant beaver habitat.

Perhaps it isn't. Yet wherever they are seen, I believe they might just have a reason for being. I drove down the road where the young couple had seen the creature hasten across the highway right in front of them. It had stood as high as the hood of their mid-sized car, and equaled its width in length. There had been about six inches of clearance below its short legs, and its color was more black than brown.

Sure enough, the road here showed a long, gradual incline from the shore, at least a mile of gentle grade, enough to accommodate tunnels well above lake levels. Perhaps the creature was out to find his own way in the world and was investigating new territory. Or, perhaps it already lived far below ground, but due to a cave-in, it dug itself out to the surface in order to get back to the lake. Then again, it may have been trying out an escape route . . . but why wouldn't it have returned to the lake via the tunnel rather than overland?

We will probably never know why it was crossing this quiet stretch of road near a Native community that knows nothing of the giant beaver—nothing at all. Giant snakes and manipogos, yes, but giant beavers are totally foreign. Yet, an hour south along the same lake is where a man dove into a tunnel near the Lake Manitoba Narrows, and another hour south of that, near Amaranth, the end of a tunnel was exposed by a cave-in where a photograph showed a clay liner in a four foot diameter void that strangely reached to within a couple of feet of the surface of the farmer's field.

That there are giant beavers living in certain areas of Lake Manitoba, I have no doubt. It's just that it is rather shocking to hear of them in places where they are least expected.

Just after Christmas 2007, I called Ross Moose, the gentleman from South Indian Lake who had told me about the big beaver and its lodge that he had seen in his trapping area. I was looking forward to going by snowmobile to see it, since a heavy snowfall the previous March had prevented it. What he told me made me thankful that we had not been able to go there. He had already been trapping in that area due to an early freeze-up, and out of curiosity, he had gone to the lake where he had seen the beaver and lodge less than ten years earlier. There was no evidence of it!

I was shocked, but not totally surprised. I had heard comments before which made me suspect that these big creatures were capable of such acts. My guide the previous summer told me his father had stated that we wouldn't find the lodge that the pair had seen in the 1980s, and someone else mentioned that if the creatures are disturbed, they will spoil the evidence of their lodge and move away.

It makes me wonder how big a threat needs to be in order to prompt the creatures to dismantle and move. Or whether they tend to move on a whim, based on conditions at the time. Ross promised to be on the lookout for any other giant beaver sites as he traveled around by snowmobile, trapping and hunting.

A new awareness of the creature as a distinct and "extinct" species is prompting some hunters and trappers in the north to watch for evidence of them now. In the fall of 2008 I met a young trapper from Nelson House whose experience I found very encouraging. Until then, I had not found individuals who had caught a huge beaver in recent years, as most stories were from the previous generation of trappers.

He matter-of-factly stated that he would get me one, since he knew where some lived. In fact, he maintained that some years ago he had caught a 160-pound beaver that his grandfather had stretched for him on a sheet of plywood, but it had not been quite big enough! He went on to say that he had seen another that was even larger. When I asked how he had caught it, he explained that he had set numerous snares along a beaver runway in typical fashion by attaching them to several poles he had put through the ice and driven into the mud. The beaver had, therefore, not only been caught around the neck by a snare—from which it could probably have escaped—but some of its legs and tail had also become entangled in other snares.

So, there is hope that a specimen may someday be forthcoming.

Although I feel I have gained a good deal of knowledge about giant beavers from the experiences and observations shared by numerous informants, I realize that there is still much to be learned. There is no end of questions, I am sure, but I would like to pose a few major ones for the sake of interest, some of which I may have alluded to earlier:

Why do they make their tunnels so long rather than just stopping well above the high water level and forming their dens there? (In several instances I am aware of tunnels being exposed a mile or more from a waterway.) Does the tunnel have any purpose other than providing shelter above the level of the water? i.e., Does it lead to a food source? Or is it one? Do giant beavers eat fish? And if they don't, what on earth do they eat in the locations where there are no muskegs nearby?

Where does all the sand go that is dug out to make the tunnels? There is a great deal of it when we consider the immense size of the tunnels compared to the ones we are more familiar with. We see the small mounds of dirt that the pocket gophers kick out, but that is nothing compared to what would come out of a big beaver tunnel. The diameter of the one hole is close to three inches, while the other is close to three feet. Where the dirt from the one hole is merely a nuisance to gardeners and farmers, the dirt from the other, if deposited on the surface, would seriously affect the landscape. At Sioux Valley, for example, where the tunnel was exposed in several spots by backhoes—half a mile from the river, and, beyond that, about another half mile—there would be over 30 semi-truck loads of sand to dispose of from the first half mile and as many again for the next half mile, not counting any branch-offs or enlarged areas. Furthermore, we do not know how many such tunnels there might be! Obviously, the sand does not reach the surface of the ground. The only reasonable place for it is in the water—in this case, a flowing river, which typically has an accelerated flow in the spring of the year.

Presumably there is not as much waste soil today as when the beavers first moved into the area and began digging tunnels. Perhaps by now they have so many expendable tunnels that they use some of them to store the sand. Why else, when we drilled 100 feet or more from a known damaged tunnel site, did we encounter soft sand rather than voids? This was uphill from the damage. Downhill from it, however, only a very short distance away, is where we encountered the void that I described earlier. So, it was evident that a portion of an undamaged tunnel network was filled with sand. What had rendered it obsolete?

Perhaps the mounds that my depth-finder showed in the river are piles of sand near tunnel openings, mounds that increase in size regularly, but get washed away for the most part as well.

How do the beavers carry the sand? I visualize a beaver bringing a mouthful (or stomach full as someone suggested), or both, to the river each time it came there from wherever it was digging, depositing it on the downstream side of the opening. Then, after it swam around for a while, looking for food, or being distracted by something unusual along the shore (like me), it returned to its tunnels, taking a mouthful, (or stomach full), or both, of construction-grade clay from somewhere deep below the river bottom, and swimming and then walking to the new construction site, plastering part of the fragile sandy tunnel before it loses its shape and begins to collapse.

Do they use their huge teeth for shaving off the sand into the elliptical shape that people describe? I would guess that would be the primary function of the six-inch-long incisors. But where these beavers build the odd lodge and dam out of trees, they obviously use their teeth to cut them down, an activity that appears to be the exception rather than the rule.

Why do they choose to tunnel into sandy soil rather than into more stable soils? I realize that the idea of these beavers making tunnels in sand is a major stumbling-block. If it is, in fact, a thick coating of clay that serves as a protective cribbing to prevent the sand from caving, that engineering feat would account for the long-term integrity of the voids. However, it would seem to be more practical for them to dig into more stable soils that wouldn't require the effort of creating liners, but such soils may not be as readily available as the sandy banks that reach up so often from different waters.

When we think of sand, and try to visualize a three-foot diameter hole in it, cave-ins would seem inevitable. The fact is, the sand that we spent so much time drilling into at Sioux Valley had a little clay mixed with it, so perhaps the creatures know instinctively what proportion of sand and clay are ideal for their purposes.

How deep under the river or lake bed do these tunnels run? To date, I have not been able to plumb an opening such as the one seen from the helicopter, but I believe these tunnels, when they first move horizontally from a shaft in the water, are deeper than we might expect. In the summer of 2007 when I was doing a good deal of drilling at Sioux Valley, a sewage lagoon was being built near the river where I suspected some beaver tunnels originated. I expected the earth-moving equipment to expose some voids, but when I spoke with the foreman after the job was completed, he assured me that they had not encountered any cavities, even though they had excavated to a depth of 12 feet.

I see two possible explanations for this. First, the tunnels may not have existed where the lagoon was built, or, secondly, the tunnels, if they do exist in the area of the lagoon, are below the 12 foot level. Where we were drilling on the side of the original river bank half a mile away, the tunnel was about ten feet below the surface. Perhaps the portion of the tunnel that crosses that distance of river flat as it connects to the river, is quite deep. Deep enough, that is, to prevent the surface water in the form of swamps and oxbow lakes from seeping into them. This means that the beavers must make certain that there is an impervious layer between their tunnels and the surface.

A rather broad question I have, which is in part just curious speculation, is which rivers and lakes might in fact contain these creatures? It would be most interesting to know. I am fairly certain that they live in the Assiniboine River, especially near Brandon, Manitoba. A corollary of this is that they may also then live in any tributaries of the Assiniboine that have high banks for them to burrow into. One tributary that qualifies is the Little Saskatchewan River, and the sighting of an unusually large, dark beaver on a bank near the village of Rivers would confirm that. The Cyprus River may be another, and I will explain why I believe so.

Remember the tunnel that was exposed in the farmyard near Somerset/Swan Lake? Well, it appears that it was only three miles from the Cyprus River, and the two are connected by a series of swamps. Not a definitive conclusion, but a reasonable one,

I think. Then again, the Pembina River is only a few miles further away than the Cyprus, but in the opposite direction, and rumors are that some tunnels were exposed near that river when the highway was built.

Only future reports of the creature will definitively determine the extent of its habitat, but based on accounts I have already referred to in this chapter, I would guess that *Castoroides ohioensis* also lives in southern Manitoba's other major river, the Red, and possibly the Winnipeg River and other tributaries that flow into Lake Winnipeg.

An experience that was shared with me in November of 2009 reinforces the theory that the Winnipeg River is home to the giant beaver. Father and son were recently hunting beavers along one of its tributaries, and, after firing a shot at something in the water, both of them watched as a large, black, bear-sized animal rushed out of the water, over a bank, and into a creek, never to be seen again. The single .22 bullet had little effect on the creature, other than causing it to exit the water.

The father told me that they were shocked at what they saw, because, although they had shot innumerable beavers in their time, this one did not match anything they had encountered before.

But it did match all the other sightings of the giant beaver.

That it lives throughout Manitoba's north is well established, I believe, so, all the evidence considered, I would think it safe to say that the giant beaver lives in much of Manitoba. And why not then also in Saskatchewan? And Ontario? And Alberta? And Quebec? I believe that once it is acknowledged that these creatures exist, there will be numerous surprises to delight us for years to come.

I would like to add a bit more information about the tunnel exposed on the Grift farm near Somerset, especially since such a large portion of it became visible. On one occasion, Ed mentioned that when they first broke into the three-foot diameter tunnel, it was "smooth and clean, like it had just been made—no smell, no debris, smooth, shiny, and hard."

The 50 feet of it that they exposed and leveled as a base for the hay shed was fairly straight and level, but then it increased in size from three feet to four feet just where it turned at almost a right angle before slanting down. (The same size that the end of the tunnel measured in the field near Amaranth where it had collapsed.)

The question of whether the big snakes inhabit giant beaver passages continues to bother me. Certainly many Native elders believe so, and since their belief is so widespread, I have no difficulty with it. However, whether or not that belief is based on very isolated observations makes me wonder. It would be ideal for large, water-based snakes to hibernate inside shafts that they could easily access from under-water, but somehow I can not readily visualize the two creatures—natural enemies, I would think, possibly competing for a common food source—co-existing habitually in the network of tunnels.

Another significant question persists in my mind, and it relates to the well driller who encountered a void at Sioux Valley. How many other drillers and excavators have had similar experiences while working on the high banks of lakes and rivers across North America? Surely Sioux Valley is not unique in being the only community where backhoes, caterpillars, and grave-diggers are rumored to have broken into these little-known, mysterious passageways.

That people and giant beavers have co-existed to this day is an intriguing thought, and the infinitesimally small percentage of humans who have been aware of them have obviously not made enough of an issue about it to make the rest of the world privy to that knowledge.

What exactly is known about the giant beavers in the scientific community? In a word—little. And those who have written about them are not fully in agreement on all the details. It is presumed that they lived in or near swamps, since that is where most of their bones have been found. In fact, the first discovery was in a peat bog in Ohio in 1837. Since then, bones of the giant beaver, or parts of their teeth, have been found as far north as Old Crow in the Yukon (north of the Arctic Circle), as well as in Alaska, southern Ontario, and in the eastern and mid-western United States.

Skeletons of the giant beaver have enabled researchers to estimate their size and weight. Length is believed to be about eight feet and the weight close to 500 pounds. Their size has been compared to the black bear, just as my present day sources have frequently done. The lack of preserved tail tissue has led some researchers to believe that its tail was round and narrow like a muskrat's, but witnesses have assured me that they are proportionately larger than the ordinary beaver tail. Ten inches wide is what one witness estimated for me, so that would seem reasonable. The big beaver that slapped its tail as I watched it made a sound very similar to the normal-sized ones I had heard so often before.

It is assumed that they ate whatever vegetation was available in the swamps. The Natives who are aware of them have not commented on what their food source is. It is entirely possible that, regardless where these creatures live, they find swamps or lake bottoms that provide them with sufficient vegetation to sustain them, yet the majority of sightings do not appear to be in that type of environment.

I do not recall anyone other than the Natives in the north mentioning that giant beavers live in tunnels. That notion seems to be as foreign to scientists as it was to me at the outset of my quest. However, since the common beavers are known to tunnel into banks on occasion, it is not an unreasonable proposition that the giants habitually do so.

Some, researchers and Natives alike, question whether these beavers ever cut down trees and built lodges, simply due to the paucity of evidence in favor of it. Furthermore, by far the majority of trappers in the north have not seen a giant beaver lodge, and many have not even heard of them simply because they and their ancestors did not happen to encounter them.

In May 2009, three years after I spotted a big beaver in the Assiniboine River west of Brandon, I had occasion to spend an evening beside the river again. I was convinced, having pondered my experience throughout those years, that if I repeated the actions that led to my sighting, I might see the big creature again. Therefore, I walked along the muddy banks for about an hour in full view of the river below. The waters had spilled over the banks in places, and were now receding gradually, leaving the silty soil barely firm enough to carry my weight, yet making it easy to see the tracks of animals. Beavers had definitely come ashore, but they were not the species that interested me. Whereas these tracks were only slight surface impressions, a creature weighing hundreds of pounds would have left large and deep holes in the soft mud.

I was hoping that I had again drawn adequate attention to myself during my walk that at least one of the creatures might make an appearance. I feel that this happened the last time as a result of my standing perfectly still for a while, so instead of repeating that action exactly, I had brought a blind with me which would allow me some movement without being seen. After I erected it, I waited patiently for some action, cameras at the ready. I assumed that the blind would pique their curiosity just like my still form had done three years ago, and I hoped that the non-threatening structure would cause at least one beaver to surface nearby so I could get a picture of it. The wind was so strong, however, that the blind was in constant motion, so once darkness became too intense, I packed up in defeat. A few ducks and geese were the extent of my sightings, and not even a single small beaver had showed itself this time.

I was disappointed, certainly, that I did not get a photograph of the giant beaver, but I knew that I had been spoiled by the awesome sighting three years earlier.

In February 2010, after the construction of a snowmobile sleigh-tent was completed, Ken Reader and I spent a weekend on the ice of the Assiniboine River where my fishfinder had indicated some unusual river-bottom configurations a few years earlier. We used the same instrument in the hope of seeing something other than fish swimming about. Being inexperienced with the workings of such technology, we were soon disappointed to find that it was not capable of showing anything other than the symbols of three sizes of fish, regardless how large an object in the water might be. Sonar was, however, the favored instrument, since the infrared underwater camera that we trained on a fish tied to a stick barely produced a picture from two feet away on account of the turbid water. Although we watched the monitor for a good part of two days, the fish was not touched by a creature of any kind, whereas the monitor of the fishfinder definitely indicated the presence of fish. There was no doubt that if the giant beaver subsisted on fish, it had plenty to feed on.

Thinking back to the gentleman who fished these waters after the Second World War, the fact that he had complained about beavers spoiling his nets perhaps showed evidence of some of their food source. His account of something grabbing his paddle and almost tipping his boat certainly confirmed the presence of something much larger than an ordinary beaver.

It became obvious to us that underwater cameras were not the answer, but rather some form of sonar imaging that could penetrate cloudy water would at least allow us to see objects and their comparative sizes. So, the search was on for the appropriate instruments, with the hope that they would not be too expensive to acquire. The benefits, we realized, would not be confined to this one site, as anything that would work satisfactorily in the river should also reveal animals in the various lakes that were known to house a multitude of unusual creatures. Equipment that was employed in the well-known waters of Loch Ness, Okanagan Lake, Lake Champlain, and others, we hoped would adequately confirm our stories.

In July 2009 I got a significant big beaver story from another unlikely quarter, and I didn't even need to leave the yard in order to hear it. I mention in the Big Birds chapter about a camper from southeastern Manitoba who shared his sighting of a large flying creature. After several hours of enjoyable discussion with Lloyd Yanz, I gave him my card which portrayed a giant beaver. That initiated a whole new phase of the discussion as he introduced his experience involving such a creature.

Campground responsibilities forgotten, I soaked up his story with delight, quietly comparing his details with other stories from the North. It was significant to me that the experience was neither set in the Canadian Shield nor in the high banks of a southern Manitoba river or lake, but rather in an isolated wilderness area not far from the American border.

Whitemouth Lake is in the Northwest Angle Provincial Forest, close to the Ontario border and Lake of the Woods, and it was near this lake that Lloyd trapped a variety of animals as a teen in the 1960s.

One fall he lost a number of beaver traps to whatever lived in an especially large beaver house, and when he used a bear trap, it disappeared as well because its chain got broken.

His mother had helped her father trap the area when she was younger, so over the winter she told him how he might capture the elusive creature. Using four #14 jump traps with teeth as well as another bear trap with teeth, he anchored them to trees with two chains, side by side. On shore he placed two extremely heavy waterlogged logs on top of the chains so that once something got caught in the traps, they would roll down into the water, drowning whatever was there.

After Lloyd dynamited the top off the large dam in order to attract the animal, the scheme was successful, and he was shocked to find a huge beaver caught by three of its legs in three different traps. When Lloyd and his brother dragged it ashore, they discovered that three of its feet were missing, quite likely chewed off in the fall when it was caught in the other traps.

The pitiful sight filled the young trapper with remorse, and his mother was similarly saddened when she saw it, and they wished they had left the big creature alone. It apparently was the only beaver that they caught there, so it may have been the lone occupant of the lodge.

The two brothers could not drag it to the truck, so they returned the next morning with a toboggan since there was still a little snow on the ground. It was all the two of them could do to wrestle it into the truck.

A Native man who skinned all of Lloyd's furs worked on this big beaver as well, and used the side of his house to stretch it on. Apparently the huge circular hide covered most of the wall and became the talk of the town.

When Lloyd went to sell it, the fur buyer refused it on account of the many scars, but offered to purchase it for its size. Instead, Lloyd gave it to his Native friend who was glad to receive it just as he had earlier been happy for the meat.

Lloyd never went trapping beaver again.

The size of the animal and the fact that the fur was black convinced me that this had to be one of the giant beavers. After all, Lloyd had said that its lodge was as big as the house he lived in, and the logs used in the dam were full-length trees.

Trappers who saw giant beaver lodges and dams knew that ordinary beavers could not have made them. It was obvious to them that some super-sized beavers were responsible, but since their knowledge of them was either very limited or nonexistent, they just assumed that some beavers had grown into extra large granddaddies which lived a very private life in these very isolated areas. Furthermore, where these structures are tucked away against banks that are covered with trees, the landscape does not tend to accentuate the size readily.

Paying heed to elders, which I believe was more common in the past than it is now, would definitely have minimized the possibility of trappers' catching giant beavers. For one thing, they were advised to avoid unusual phenomena. Those who did not pay heed, and, innocently or otherwise, set traps and snares where these creatures lived, lost many of them. So between the unaffordable reality of losing precious traps, and the mysteries attached to the unusual structures, trappers would have been inclined to leave well enough alone. The few who managed to catch some of the large juveniles knew from their color and immaturity that there was something out of the ordinary, and it was some of these trappers who shared with me their perplexing experiences. Whether or not they believed my explanation of a separate and distinct species of beaver I am not certain, but the thing that I am certain of is their interest in discussing a topic that has long puzzled them.

As I said earlier, some Native traditions in Canada and the United States apparently include stories of the giant beaver. Jane C. Beck, in her 1972 article entitled "The Giant Beaver: A Prehistoric Memory?" identifies a number of tribal areas where these stories prevailed. She mentions the Northeastern Algonkian Indians, Montagnais, Wabanaki, Malecite, Micmac, Passamaquoddy, and Ojibwa, and to the west, the Beaver Indians of the Peace River region of British Columbia. Beck states "Alfred Romer tells us that *Castoroides ohioensis* was common in the northeast, and absent only in the southwest. (Romer 1933:53) However, it appears from fossil evidence that the giant beaver was most abundant in the region of Indiana (Cohn 1932:237). Remains have been found throughout the central part of the United States and along the Atlantic coast from Florida to New York. In the far West its remains are rare, but a single occurrence in Oregon shows that it did cross the Rockies (Simpson 1930:311)."

Believing that the incidence of folktales involving the giant beaver would indicate "a remnant of a former memory—the memory of the Pleistocene beaver," Beck goes on to say "if the mammoth was fossilized in folktales, why not the giant beaver?"

I, of course, based on my own experience as well as those of countless others, would go one step further and say that if the giant beaver is fossilized in folk tales, is it not that much easier to believe that it is still with us today?

In Manitoba's northern wilderness, this unique but ordinary-sized beaver lodge covered with limestone rocks was photographed by Brian Buchberger of The Pas.

MANIPOGO

The name Manipogo seems to be used for any unusual, unknown creature that is seen in Lake Manitoba. The fact is, however, that descriptions vary so widely that it seems evident that there are a number of large aquatic mystery animals that not only share this large lake but also the name of Manipogo.

The Native people who have lived along its shores and used it as a source of food and transportation for centuries have been fully aware of the mysteries of the lake, and when Europeans moved into the area, they soon made their own discoveries. Although some of the early accounts go back about a hundred years, it was in the 1960s that there was a rash of sightings in the area now aptly named "Manipogo Beach" on the west shore of Lake Manitoba, and the name Manipogo, an offshoot of British Columbia's Ogopogo, has been used ever since.

Many reports and rumors of water creatures of all shapes and sizes but too numerous to mention or investigate, have been made by folks living on both sides of Lake Manitoba. I have chosen here to report mainly those stories that I heard from the eyewitnesses themselves, or from people who knew them.

I have arranged the creatures into categories, and, for my own purposes, named most of them after the people who gave me the first or clearest descriptions of them. The first two, Bina and Vera, have the first names of a mother and daughter, respectively; the third is the first four letters of the last name of a gentleman by the name of Flamand from St. Ambroise; the fourth is the first four letters of the last name of a gentleman from Lake Manitoba by the name of Maytwayashing; the fifth is the first four letters of the last name of several gentlemen by the name of Traverse from Lake St. Martin and Fairford; the sixth I call Swan which is the last name of another gentleman from the Lake Manitoba First Nation; the seventh I call Loopy because of its tendency to pull itself together into loops. In addition to these, Giant Snakes, Giant Beaver, Water Dogs, and Beaver Ducks are also believed to inhabit Lake Manitoba. This list of creatures may only represent a portion of what is really out there, but it is a tantalizing assembly nonetheless.

STRANGE CREATURES SELDOM SEEN

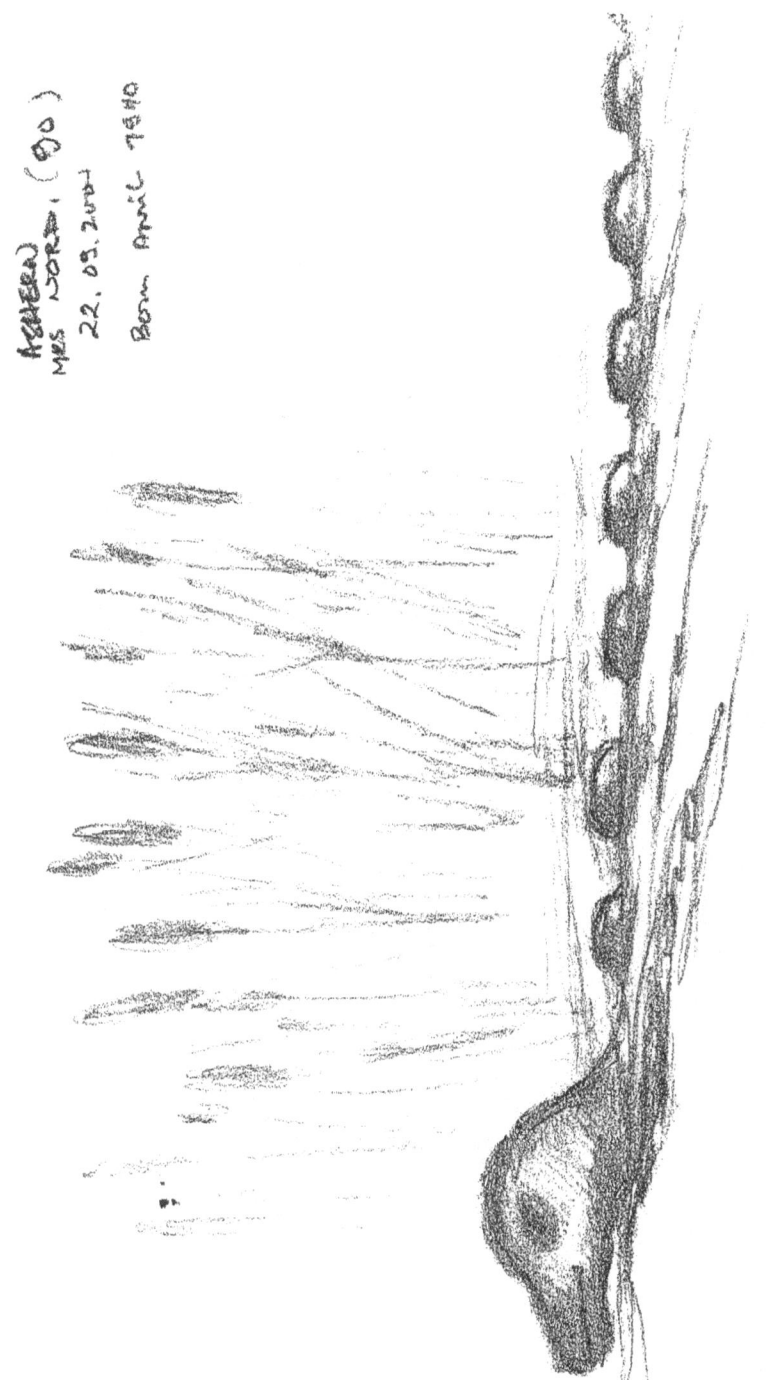

Seen by Bina Nord in Lake Manitoba and described to artist Jarmo Sinisalo.

MANIPOGO NUMBER ONE: BINA

BINA NORD WAS LIVING on Peonan Point—a long peninsula at the north end of Lake Manitoba—in the 1950s, where she and her husband farmed. He and their young children had gone across Portage Bay to Steep Rock by boat one day, as this was their closest post office and shopping center. Bina told me that she walked down to the lake to fetch a pail of water, and noticed something among the reeds near the shore. At first glance she thought it was a deer's head just above the water, but quickly she realized that it could not be, as it was dark and shiny, and had no ears. When it saw her, it swam away, revealing about a dozen bumps on its long serpentine back. Each "bump" she estimated to be about a foot high, with about a foot in between them. An artist sketched the accompanying picture in her presence and to her satisfaction. When I showed it to others around the province, it was recognized by very few first-hand observers, but it reminded a few people of what they had heard from others, the numerous small bumps matching some descriptions.

The head of a deer and that of a sheep are somewhat similar, and the latter description was recorded by the Winnipeg *Free Press* in 1957. Three years earlier, two resident fishermen out in a skiff encountered the creature near where they lived at the north end of Lake Manitoba. They apparently only saw the head and shoulders of it. (The same article mentions the 1948 sighting by the area's first teacher, C. P. Alarie, but probably of a different creature, according to the description I recorded in the crocodile chapter.)

MANIPOGO NUMBER TWO: VERA

BINA'S FAMILY HAD DIFFICULTY believing her when they returned home. In fact, one daughter, Vera, told me she was skeptical about her mother's sighting for about 40 years, when finally she saw her own creature in the lake.

Being an avid fisher, Vera would often be angling off the rocks that Steep Rock is so famous for, directly across the lake from where her mother reported her sighting. She was sitting there one calm, sunny day when she sensed something beside her in the deep water. Turning her head, she saw a large, barrel-shaped form rolling down into the water out of sight. It swam away, displacing the water as only a large creature could do. She saw its size, color and pattern well enough to describe it to the artist.

It was after this experience that she began to believe her mother's story, for she now had one of her own.

What the rest of the creature looked like, we do not know. The head is seldom, if ever, seen. It may not even be the same creature that strikes fear into boaters when they see a large round hump moving along in the lake.

STRANGE CREATURES SELDOM SEEN

Described by Vera Nord, and painted by Jarmo Sinisalo, this large serpentine creature was seen near Steep Rock on Lake Manitoba.

One man from St. Ambroise, (Flam), described to me his terrifying experience as a young man. The beach on Lake Manitoba was just a couple of miles from where he lived, so, as his habit was, he would go for a swim when he got home from work, and wash his hair as well.

This one day, as he was soaping up, he caught a glimpse of something a good distance out in the lake. A few seconds later, after he had rinsed off the soap, he looked again, and to his horror, he could see the back of some large creature much nearer. Knowing that it was coming towards him at a great speed, he swam madly to shore, expecting to be grabbed from behind at any moment. He did not stop swimming until his hands touched the shore, and even as he sprinted to his car, he half expected something to follow him. He had not seen a head, but what he did see told him that the creature was of a considerable size.

It may or may not have been the same as what Vera saw, but what he saw 26 years later was quite likely something altogether different.

MANIPOGO NUMBER THREE: FLAM

IT WAS IN THE SUMMER OF 1991 that Marcel Flamand took some of his children and a few relatives by horse and wagon to a spot between St. Ambroise Beach and St. Laurent. The children were in the water, watched over by his niece. Suddenly she saw some large, dark forms rise a few feet out of the water just behind where the children were, so she shouted to the men who were nearby. Marcel told me that, from the tone of her voice, he knew instinctively what was happening, so the children were quickly hustled ashore.

As if that hadn't caused enough excitement already, they saw, as they sat on the beach gazing out on the lake, a creature arch its long neck into the air, pause for a moment, and then plunge back into the water. A few minutes later, it happened again in a different spot. And yet again. Five or six times the creature, or creatures, "put on a show," but never were two visible at the same time. The creature's neck was estimated to be about 20 feet long, and the head had a proud and majestic appearance "like an Arabian horse," only a little smaller. Whether or not this was the same creature seen near the swimming children is not certain, but it seems likely.

Here again are examples where water activities that involve movement and splashing tend to attract creatures. Swimming, washing hair, children playing . . . and yet there don't seem to be reports of people being attacked, even though there would seem to be many opportunities for that to happen. The streaker from Lake St. Martin (mentioned in the Big Snakes chapter) is one vivid example!

The only other account I heard involving a long-necked creature of this sort was from Genevieve Stagg of Fairford, who told me what she had seen in Lake Manitoba

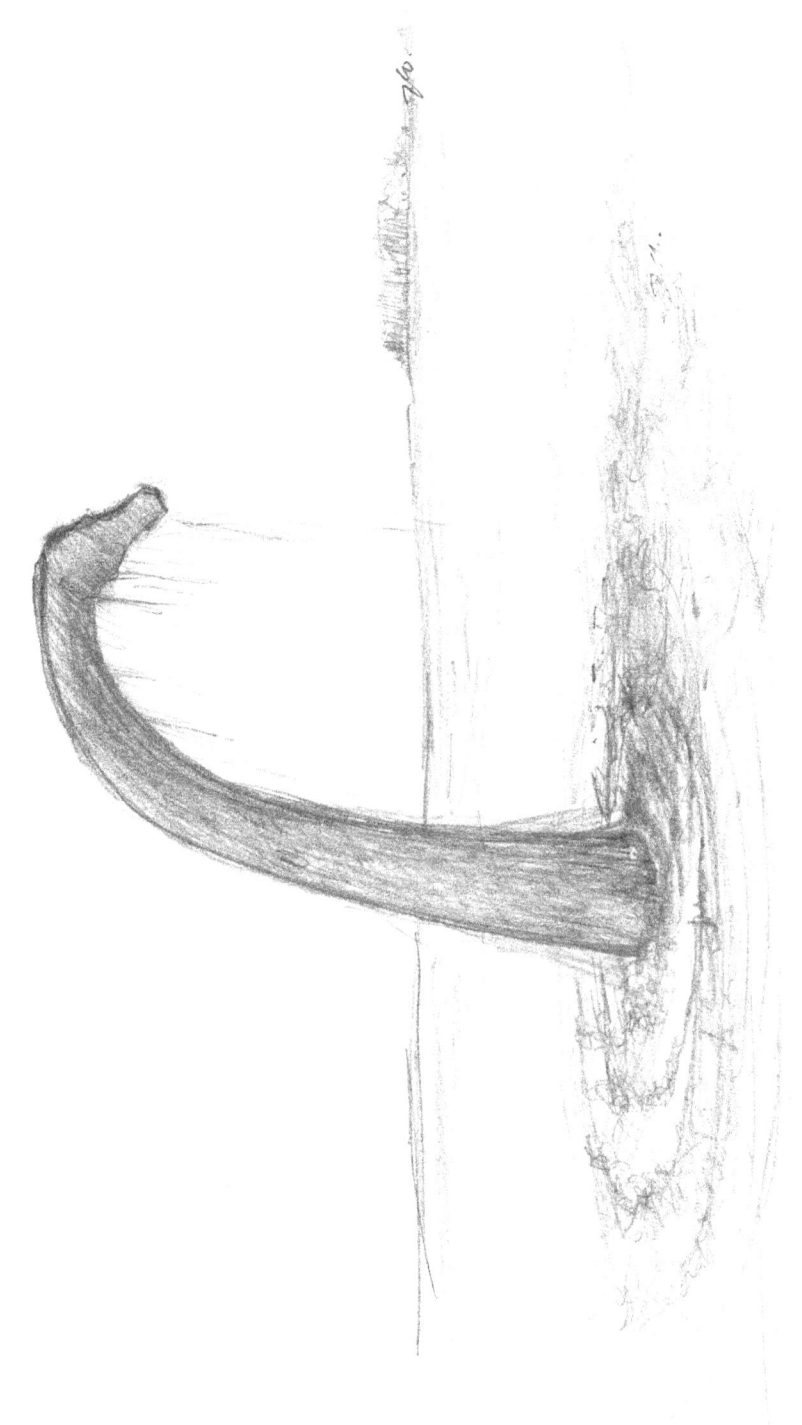

Long-necked creature witnessed by Marcel Flamand and others in Lake Manitoba near St. Ambroise.

at Crane River one summer. She, as a teen, had gone to visit her sister there, and had been standing with some others near the lake on Treaty Day. She was the only one of her group who was facing the lake, and suddenly she saw a long neck rise out of the water, pause briefly, and dive back out of sight. She had been so shocked that she couldn't speak until after the creature had vanished.

Later, when there was an opportunity to cross the bay by boat, she refused, choosing rather to walk several miles around by the bridge.

MANIPOGO NUMBER FOUR: MAYT

TWO MEN FROM THE LAKE MANITOBA FIRST NATION near the Narrows saw what they thought was a horse swimming across the lake. They soon realized that it wasn't one because the head was too shiny, and it had no ears. About 18 to 20 feet of something was visible, and they were able to watch it for at least 15 minutes from less than 50 yards. Its head and neck were about the size of a five-gallon pail (close to a foot). The head reminded them of a horse's, but it was bigger than one, with frog-like transparent eyelids that blinked occasionally. It was a lime-green color, with yellowish spots. When it submerged, it did not dive, but went straight down.

One of the men was inclined to go public with the story, but the other, John Maytwayashing, who shared the sighting with me, did not want to become the laughing stock of the community, so they agreed to keep the incident to themselves.

A few miles away and some years earlier, a similar sighting took place, but on a regular basis. Different folks tell of an old man, Andrew Sumner, who used to go out from the beach area in his rowboat and lift his nets each morning. The fish he did not care for he would throw out to some creature that was often there, and he was known to make the comment that it was harmless. It apparently was at least 30 feet long and had a big head. It is entirely possible that these two stories are of the selfsame creature. The old man's grandson corroborated the story from his own memories, and grandfather had even sketched the creature for him.

It may have been the same one that some other fishermen encountered in the same general area years later. Only they didn't have the same respect for the creature as the old man had. When they had gone to lift their nets, a large head had appeared, and when the men took off with their motorboat in terror, the open-mouthed monster had chased them until it fell behind. When they arrived on shore, pale and shaken, they related their encounter to someone whose comment was, "It was angry because you were taking away its breakfast."

A similar story is told by some on the Pine Creek First Nation, near the south end of Lake Winnipegosis. (The two lakes, Manitoba and Winnipegosis, are joined by rivers, so the "Manipogos" can presumably go back and forth if they so desire in

Lake Manitoba creature witnessed and
sketched by John Maytwayashing (Mayt)

Manipogo experienced by other men in a boat.

times of high water). Again it is about an old man who went out with his rowboat to lift his nets. A big head broke the water, scaring him so badly that he rowed back to shore in record time. Someone who had observed him from a distance wondered if he had acquired an outboard motor. These two stories have been a source of amusement in the two communities for quite some time, and will likely continue to be, as they are passed down through the generations.

In 1992, two boats were in the bay at Manipogo Beach, a husband and wife fishing from one, and two men from the other. All were from Portage la Prairie, although they did not realize it until later. When they stopped their motors, they heard, in the evening stillness, the sound of turbulent water, and saw a creature with its large head out of the water, moving forward at a rate that produced a big noisy wave before it. I had the privilege of speaking with both parties, who some years earlier had their stories recorded in the local newspaper.

Off the northwest shore of Peonan Point, a man watched a large form swim away from him, and as it went, he could see fish jumping out of the water on both sides of it.

A man from Vogar, near Lake Manitoba First Nation, told me that he and several others were out in a boat when they saw, from less than 100 yards away, a creature with a big head swimming towards them, its mouth open. He said they had been very happy to have a motorboat to escape with.

None of these stories include any details that would indicate the method of propulsion, but the assumption generally seems to be that flippers are a reasonable guess. If legs were an option, then we should see the animals and their tracks on land. Flippers would account for the powerful thrust needed to push their large bodies through the water.

One sighting in the Berens River some years ago may shed some light on this. A couple had noticed something large swimming in the river where there was swift water, and the creature had crossed it without any noticeable setbacks resulting from the current. After a while, it had swum to shore where the rock bank sloped steeply into the river. The creature had made as if to climb ashore, pawing the rock with what appeared to be flippers on the fore part of its body. This had lasted only a short while before it swam away. Its moose-sized head makes it a match for Mayt the Manipogo.

MANIPOGO NUMBER FIVE: TRAV

ALTHOUGH THIS CREATURE seems to reside in Lake St. Martin, I will include it with the Manipogos because of the lake's proximity to Lake Manitoba where something similar has been reported. It may very well be the same creature that is seen in some other lakes as well. Lake St. Martin is not a likely lake for an animal that is often described as being the size of a horse or cow. Although it stretches for 30 miles, it is a fairly shallow lake. Nevertheless, a variety of sights and sounds seem to indicate that something besides the large snakes lives there, something that seldom gives evidence of its presence.

An early report that is often recalled comes from a sailor who enjoyed traveling between the three communities of Fairford, Little Saskatchewan, and Lake St. Martin in his sailboat. He told of meeting a large form that had risen out of the water in the darkness near his boat.

A man from Fairford described to me what he had seen in the lake as a boy. He and his older sister were returning home after lifting a net in small Pineimuta Lake that is connected to Lake St. Martin. When he looked back, there, out in the middle of the lake, appeared what looked like the chief's "Holstein-colored" horse. When he had pointed it out to his sister, she had just brushed it off by saying that horses don't go into a lake. All his life he had wondered about what he saw, as few such reports were able to corroborate his.

It was in the 1970s that the most definitive sighting took place just south of Dunsekikan Island in Lake St. Martin. A boatload of men were spotlighting for deer or moose when their beam fell upon a large form in the water. The body seemed to be about the size of a cow's, but the neck stretched four or five feet out of the water, making it obvious that this was not an identifiable creature. Although the head was somewhat similar to that of an elk, there were too many details that pointed towards it being a totally unique creature that lived only in the water. For example, after the light had shone on it for a few seconds, the animal submerged, never to be seen again.

The men were so rattled from the experience that when they set up camp for the night, they chose a high bank well away from the water.

Others who report something unusual in the lake often mention seeing two giraffe-like horns above the surface, and in some cases, the top of a wide head is seen as well, with the eyes set far apart.

Perhaps this is the same creature that the driver of a boat loaded with people saw in Lake Manitoba when he was setting out for Peonan Point from the Fairford River in the middle of the last century. Barely into the journey, he had asked for the axe, and immediately headed in to shore to set up camp. It was only after the trip was over that he acknowledged seeing two stubby horns protruding from the water in an area where no stones were near enough to the surface to support two birds standing side by side as he had initially thought.

This creature, which has been known to bellow like a cow, tends to rise far out of the water after dark, or just at dusk.

Near Moose Lake, east of The Pas, a lake called Cow Lake by the Natives seems to have the same reputation as Lake St. Martin, according to stories of a cow-like creature. Hunters, I understand, tend to avoid the smelly lake with its questionable inhabitant.

The animal that I refer to in the Big Birds chapter, called a water buffalo by some folks in the Waterhen area, may be different from both Trav and the Underwater Moose. From the few eyewitness descriptions of the creature in Lake St. Martin and Pineimuta Lake, however, it does not appear to be the underwater moose.

MANIPOGO NUMBER SIX: SWAN

A YOUNG MAN BY THIS NAME described seeing the back of a creature in the waters of Lake Manitoba. What stood out in his mind, and which he sketched for me, was a low row of disks in the centre of the back. This seems similar to what Charlie Burrell had reported to Oscar Fredrickson from in Lake Winnipegosis. (See the story in the Whale chapter.)

Creature in water at the base of cliff at Steep Rock. Jamie Bittner sketched what she recalled seeing as a young girl while fishing off the cliff.

What Mr. Fredrickson himself saw in the form of ice slivers pushed up by something with protrusions of some sort on its back may possibly be the same creature. He says "There was only a narrow row of icicles pushed up. Something with a big fin or a disk on its back must have gone under the ice in order to push up the slivers of ice in the manner this did."

For those who are not familiar with the break-up of lake ice, Mr. Fredrickson's description may be difficult to understand. Let me elaborate briefly to explain what he most likely saw. Ice is generally viewed as a solid sheet floating on top of the water. With the advent of mild spring weather, however, the air and water temperatures rise, softening the ice. As the meltwater from the surface finally drains through the rotten ice, it creates small openings leaving what appears to be a mass of icicles side by side in a vertical position. When these icicles are melted to the point where they are barely attached to each other, anything swimming under the ice and touching it will push up a contour that matches the shape of its back. It was this contour that I believe Oscar Fredrickson witnessed as something large and powerful "with a big fin or disk on its back . . ." passed beneath the cake of ice in front of him.

In the same manner, witnesses have reported seeing *solid* ice being pushed up as something obviously massive and powerful attempted to rise to the surface. Perhaps such events were occasioned by a desperate need for air, which thought returns us to the question of where these creatures obtain air to breathe on a regular basis.

The only clue I am aware of to date hails from Wilson Moore of Nelson House who remembered seeing small, dome-shaped configurations on the ice around open holes. Some bits of vegetation and debris surround the hole, just as if something had spit or breathed it out when coming up for air. No tracks were visible around the holes, so Wilson had naturally assumed that they were "blowholes" maintained by some aquatic animal.

There is also the possibility that Bina's creature with its foot-high bumps might be the one that lifted the slivers of ice in Lake Winnipegosis, even though she saw it in Lake Manitoba. With the two lakes being connected, there is a good possibility that this species lives in both lakes. As for the whales, I have not heard such creatures described from in Lake Manitoba, so perhaps they are exclusive to Lake Winnipegosis.

MANIPOGO NUMBER SEVEN: LOOPY

A MAN FROM A COMMUNITY on the west shore of Lake Winnipegosis thought at first he and his companions were seeing three tires standing up in a row. Other boaters from the community hit something with their motor, and it swam away with a vertically undulating motion.

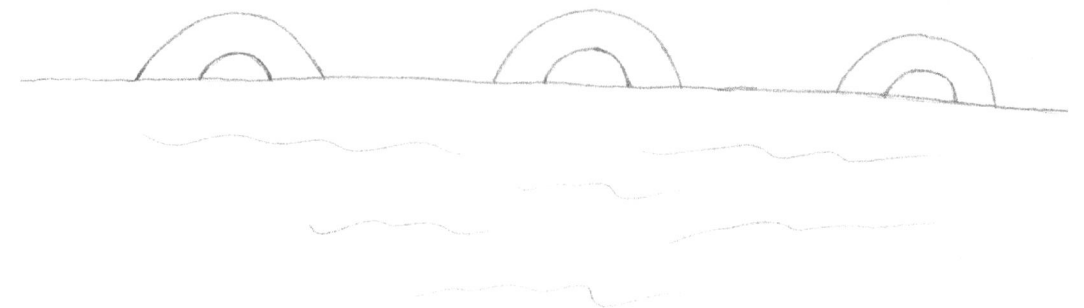

Loopy, a phenomenon of numerous lakes around Manitoba.

Snakes move solely in a horizontal, weaving fashion, so whatever these creatures are, they are not true serpents. Since they do not seem to come out of the water, their full size and characteristics are very much in question, but their ability to pull themselves together into one, three, and occasionally two loops is legendary in lakes throughout Manitoba. The diameter is typically estimated at about a foot.

In the summer of 1962 a rare photograph of "Manipogo" was taken in Lake Manitoba when three fishermen saw "a large black snake or eel that swam with a ripple action." Its diameter was about a foot, which is similar to the many reports I have heard. The photo shows only one loop which is somewhat more elongated than the typical "tire" descriptions, but there are definite similarities.

A Winnipegosis fisherman, Ralph Sanderson, had his family with him in the boat one day when they spotted something like a log in the water. According to the daughter, Karen, who I spoke with recently, her dad had stopped the boat to clean weeds from the propeller, and during that time the object had been seen a short distance away. When her dad had started up the motor and headed in its direction, they saw it loop into two see-through round shapes before submerging. Her mother had grabbed a paddle and stood up in the boat, something that she never did (and never heard the end of), and as they approached the spot where the creature had been seen, there was only disturbed seaweed visible. The next day her dad heard several reports that nets set in that area had holes in them.

Although the following story is probably not about Loopy, and perhaps not even about any of the other Manipogos mentioned here, I will include it here.

Karen's younger brother Rick, who was with the family and remembers the incident, experienced a similar one in the same lake about ten years later when he was out commercial fishing with an older cousin, Alvin Thompson, who owned the 47-foot *Barbara Ann*. When they were stopped, perhaps for lunch, they saw something in the water not far away. Alvin recalls seeing a large object about a foot out of the water, with a jagged fin-like growth on top, perhaps 18 inches long. From his vantage point in the wheelhouse, he knew that this was not just a log, but something he had never encountered before, and, curious as to what it might be, he started the diesel engine and slowly moved towards the object which quickly began showing signs of life. As he approached it, and almost ran over it, it submerged, but remained visible to the men who were now watching it from the bow of the boat. Rick described what he saw as being about a foot in diameter, and since it was a calm, sunny day, he could see down the four or five feet and recognize the color and pattern that reminded him of a jackfish. He could see that the creature was longer than their boat, and recalls his heart pounding from the excitement. One of the men suggested going below for a gun to shoot it, but Rick had not liked the idea, fearing what an injured monster of that size might do.

After a while, it surfaced a short distance away, so Alvin decided to follow it. It kept well ahead of him, even after he opened the throttle, achieving around 11 miles per hour. The turbulence behind the creature indicated to him that this was no small animal, since the wake was similar to that made by a ten horsepower outboard pushing two people in a boat. The method of propulsion was not obvious to any of the viewers, so it is not known if it had flippers or not. It did, however, swim at the surface where it was visible for over a mile of the chase before it finally submerged and vanished. Alvin is convinced that the front portion of the animal, which no one seemed to see clearly, had to be much larger than the part that his cousin saw.

Although their story brought on a good deal of laughter when they told it at the fish shed, they were vindicated by reports of holes in nets in the following days.

Alvin stated that they were not the only ones to report such a sighting. He knew of a couple who had followed a large creature out the channel from the Mossey River into the lake, and it had stirred up mud even though the depth would have been close to a dozen feet. His grandparents also, with a 15-horsepower engine, had once followed a large creature.

Most of these sightings took place in the sixties and seventies, after Oscar Fredrickson's time, but they were made by people he would have known from the town of Winnipegosis. It might have been enlightening to have him use his knowledge and experiences to compare their stories with his, as there were some similarities.

At Kinoosao, on Reindeer Lake, children reported seeing something in the form of three loops, and when it submerged, another surfaced nearby.

At Nelson House, when an elderly trapper saw the sketch of the three-looped creature, he remarked that it was similar to bloodsuckers (lampreys) he was familiar with. He said they could be up to several feet long and sometimes bunch up in the same way. That raises the question whether lampreys really get that big, or whether this is something else.

Two fishermen setting nets on Wuskwatum Lake west of Thompson reported seeing Loopy as well, and in Leftrook and Highrock Lakes northeast of Nelson House, something that moves in an undulating fashion was seen there.

A man and his wife from Pukatawagan saw a "bloodsucker" swim alongside their canoe, and it was longer than their craft. It was stated that such creatures don't attack people. Foot-long bloodsuckers seem to be common there.

In shallow water along the shore of Spruce Lake east of Lake St. Martin, the grandfather of Gordon Stagg Jr. of Fairford had mentioned seeing a big, brownish "bloodsucker" that was over four inches wide, and between four and five feet long. Lampreys several feet in length have been reported caught in nets in Lake Winnipeg.

Whether or not it is just a larger version of the same creature, mention is made of a huge, tubular body in Lake Manitoba, possibly similar to the one seen by Vera. A man from Fairford looked down at the lake when he was crossing the bridge at the Lake Manitoba Narrows, and was shocked to see near the surface a long, serpentine body that he estimated to be about two and a half feet wide and about 40 feet long. It was "like a giant leech." Several fishing boats were nearby.

Another man recounted his experience on Lake Manitoba as a boy. His grandparents had a cottage at St. Laurent, so he got to fish there occasionally. On one trip, while their boat was anchored, and everyone's lines were in the water, a "hump" rose slowly above the surface of the water about 25 feet away, flowing soundlessly from right to left, rising until it cleared the water altogether in the center. As it flowed, the "barrel-sized' object gradually tapered smaller, and sank out of sight. It was smooth, shiny, and beige in color, with no scales or texture. It submerged without exposing its tail.

Its tubular, see-through loop makes it similar to other accounts of the Loopy, but its huge size and beige color differentiate it markedly from them.

The size compares more with Vera's, but the distinctive pattern that she described is lacking.

A rather unusual account from Lake Winnipeg that has some obvious similarities to the long, serpentine Manipogos, would also fit into the Lake Winnipeg Monster section. It came from a fisherman from Sagkeeng, a First Nations community on the banks of the Winnipeg River where it flows into Lake Winnipeg. He had gone out in his boat to lift some nets when he saw a long, grayish form with a tapered tail slide into the lake from on a rocky island. He and his partner had just the day before, after they had set two nets nearby, climbed onto the island to erect a marine marker that had fallen over, and it was around this marker that the creature had been coiled. He found his nets in a terrible tangle, and instinctively made the connection. He kept a fearful eye on the water around him as he tried to extricate his torn and empty nets, hoping that he would not see anything. Shortly, however, a large, tubular, grey form appeared above the water directly in front of him, looking for all the world like the funnel on a ship, and it turned its "head" towards him for about ten seconds. Since I had not realized from my first visit with him just how close the creature had been, I was surprised when I recently had the opportunity to ask him about that distance as we stood near each other on the porch steps. "From me to you," was his reply, and again he explained that the shock had been so great that he had been quite ill for months. He had managed to get home that fateful day, but he had never resumed his occupation as a fisherman again.

I thought that I was finally going to hear a description of a "hydro pole's" head, but I was disappointed when he stated that the face of the thing reminded him of a blood sucker!

The island on which this 50- to 60-foot creature had been seen resting was named "Devil's Island" by the locals, and we can now understand why.

THE ABOVE MANIPOGO STORIES are just some of the many sightings that likely number in the hundreds. Only time will tell which other "Manipogos" need to be added to this list. For example, the creatures that I have discussed in the "Loopy" section might be any number of distinct animals that in time will warrant treatment as separate species altogether. The creatures described by Oscar Fredrickson in the "Whale" chapter also reside in the same water systems with the rest of the "Manipogos."

It is difficult to imagine such a variety of creatures cohabiting such a relatively shallow body of water. It is even more difficult to consider in what various ways the individual species, not to mention the members within each species, survive the winters under many feet of ice. The ones that have the capability of creating their own tunnels, like the giant beaver, are not a conceptual problem, but those that have not that ability must cope in other ways. The possibility of cracks and caverns accessible from underwater has been mentioned elsewhere as likely hibernacula, and this idea seems to make good sense. The view that some Manipogos perhaps swim to the ocean at Hudson's Bay via the various rivers and lakes is no longer entertainable, as a dam on the Fairford River would preclude that possibility. If reaching the ocean were a requirement for survival, then there should have been a mass kill-off of desperate creatures in the last half century since the installation of dams, and their bodies would have littered the shores. Furthermore, from before that time, there appears to be no history of fall migrations which would have been

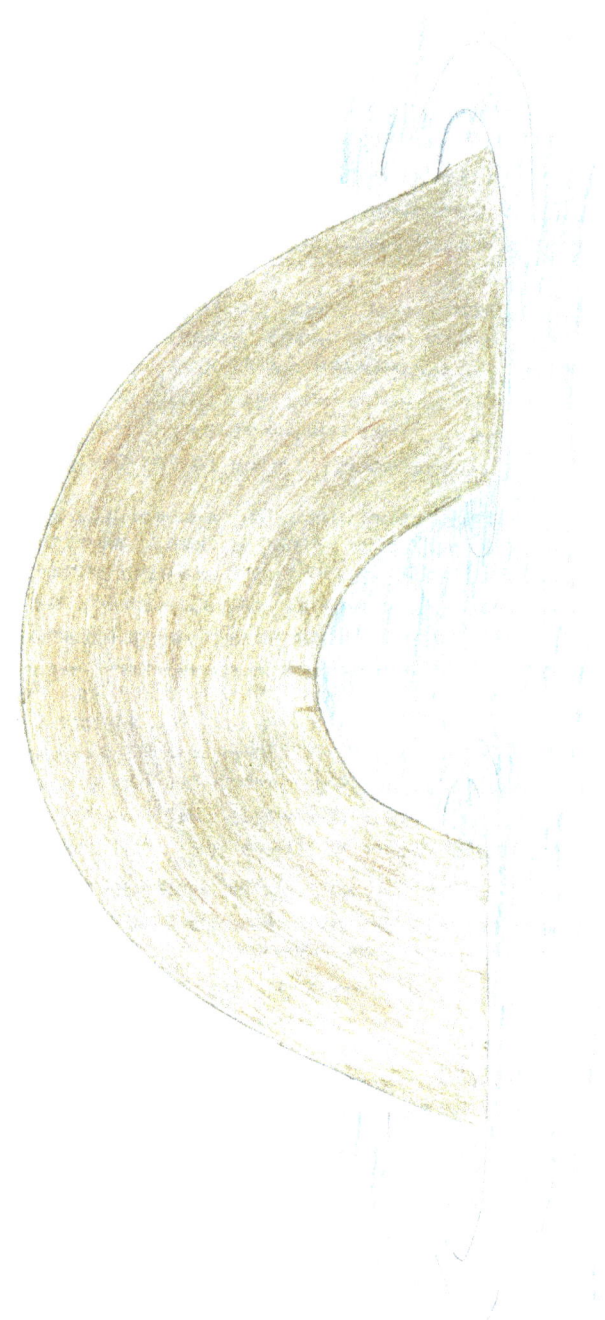

One of the Manipogos seen near St. Laurent and sketched by witness Rob Cormack.

witnessed by at least some residents during times of low water when large creatures would have had difficulty navigating not only the shallows of the Fairford River, but also those of the Waterhen River systems that connect Lake Winnipegosis to Lake Manitoba, and the Dauphin River which leads to Lake Winnipeg. Sightings continue, however, to this present day, providing ongoing evidence of unidentifiable marine animals that obviously cope with our winters.

There are few reports of broken nets among the winter fishers on both Lakes Winnipegosis and Manitoba, so it would follow that there must be very little activity by its denizens during that time of year.

BIG SNAKES

IF IT WAS THE RUMOR OF SASQUATCH SIGHTINGS in the 1970s that first brought to the attention of this farm boy-turned-teacher-gone-north the possible existence of unusual creatures, then it was the rumor of a large snake sighting in Manitoba's Interlake in the late 1980s that jolted my imagination and spurred me into action. I resolved to determine the extent of knowledge in this realm, and began by questioning experts in the field of zoology, and more specifically, herpetology—which is the study of reptiles and amphibians.

To my surprise, the consensus seemed to be that the garter was Manitoba's biggest snake among a very small handful of species. (Only five species are recognized: the eastern garter snake, the plains garter snake, the red-bellied snake, the western hognose snake, and the smooth green snake.) When I introduced the possibility of larger snakes existing here, our climate was always cited as being far too harsh for them to survive. Now, over a dozen years later, nothing has changed, except that I have heard from hundreds of down-to-earth people that there are a variety of small, medium, large, and extra-large serpents that share our turf.

Many who experienced an encounter wish they hadn't. Fear of snakes is a common phobia, one with which I still struggle, having grown up with an intense hatred of the creatures. Anything between shotguns and hammers was adequate ammunition to rid my space of them, but, as the positive contributions of the species began to stack up over my lifetime, I developed enough respect for them that I have taken to driving around them instead of over them. And, because of the multitude of diverse sightings listed in my notebooks, I have even developed a fascination for them. Someday I may even touch one—if there is a definite need to do so.

I can clarify something about a large snake shed whose appearance caused quite a stir almost a decade ago. Newspapers covered the various theories that resulted, including some from experienced professionals. Not knowing anything about snakes, but having heard quite a few giant snake stories by 2001, I was hoping the shed would prove that Manitoba does indeed have super-sized serpents. When I found someone who was willing and able to do the DNA testing on a piece of the shed, the results indicated what the professionals had said all along—that it was from a *Boa constrictor*. Brian Crother, of Southern Louisiana University, found that the specimen matched most closely the boas native to northeastern South America. The consensus in Manitoba had always been that if, in fact, the shed was that of a large serpent living in Lake St. Martin, then, being a tropical snake, it could not have lived there the previous winter, but must have either escaped from captivity, or been released from it. Regardless of how the boa shed ended up in Manitoba, the incident certainly engaged the interest of folks around the province, creating a greater awareness of giant snake stories.

I live across a sizeable wilderness from where a giant snake had been seen crossing the highway on the Peguis Reservation. Only lakes, rivers and forest separate us, so I wondered why I hadn't heard about such creatures where I live at the northwest tip of Manitoba's Interlake. I soon discovered that the reason was simple—I hadn't asked. And once I began enquiring of the locals, and then eventually further afield, there was no limit. Stories were available for the asking, and references multiplied exponentially.

Why then, you ask, if there have been so many sightings, don't we know more about them? That is definitely a valid question, and one I have pondered for years. Perhaps I have, in part, given one reason already, but let me explain.

It is human nature to be skeptical about anything new or unusual. It is also human nature to let the bearers of unusual stories know what is thought of them. So, it doesn't take an individual with such information long to determine what the safest and most comfortable approach should be.

This fear-of-being-laughed-at factor is likely the single most significant reason why we don't hear many stories. However, with the media in recent years thriving on the extraordinary and the bizarre, it is slowly becoming more acceptable to be an active participant—and many witnesses who hear of others' experiences and then share their own have acknowledged the cathartic effect it has. They seem much reassured that they are not unique in witnessing something unusual.

Another reason for the paucity of reports is rooted in the culture of the majority of the witnesses. The Native peoples, who have generally lived by the waters in somewhat isolated communities for generations, are by far the richest source of information that I have, but because of ancient, deep-seated cultural taboos, some are still reluctant to delve into the subject for fear of recrimination. The advice of the elders has long been to give anything mysterious a wide berth so that nothing harmful might befall.

The lady who had seen the large snake on the highway in Peguis had no qualms about telling me what she saw. In fact, after she was finished relating her experience, she referred me to several others in the community who had had unusual snake encounters—and the ball has not stopped rolling since. Here then is the account from the woman who I determined was well-respected, industrious, and credible.

Winnie was almost home from a trip to Winnipeg in her minivan when on the pavement ahead of her she saw what she thought was a big log that someone had inadvertently dropped in the middle of the road. She slowed down to drive around it, and as she did so, it moved. In terror, she sped off, realizing that it was a huge snake. When others went to survey the area, they found the evidence in the grass.

From the same community comes an account of hunters going after ducks on a lake by moonlight, and firing at what they thought was only a handful of ducks sitting on the water. The turbulence that resulted indicated to the hunters that they had hit something quite different from the intended target.

Also from Peguis comes the following which I hesitate to share, and would not, if it had not come firsthand from a respected elder. In good faith he shared his experience with me, and I will relate it as I recall it, relying on my memory which is reasonably trustworthy in matters that are important to me. I cannot refer to my notes on much of my early research, as they were destroyed in a house fire in 2002. And, keep in mind that I am only the tale-bearer, and have to deal with my own skepticisms.

As a young man, Michael had gone duck hunting on foot with a friend. They skirted the edge of a lake, hearing an unusual call as they went along. The sound was repeated now and then, and seemed to keep pace with them, but neither of the men acknowledged any discomfort on its account even though it was an unfamiliar, eerie sound. When they reached a swampy area, they decided to retrace their steps, and immediately came upon a huge serpent, its mouth wide open, emitting the fearsome sound that they had heard before. Whether or not they fired a shot at it before they fled I don't recall, but I do remember the gentleman saying that they would have easily outrun an ATV on that occasion.

Another man from that community who had heard this story firsthand like I did said the sound that the serpent made had been compared to a woman screaming.

This story reminds me of a similar scenario set in the nearby central Manitoba wilderness north of Gypsumville where there is also nothing but swamp, lakes, streams and forest. A man, I was told, was walking by himself along a lakeshore in a reedy area when he saw a large serpent-like creature. It gave him such a fright that he lost full consciousness, only to regain it some time later after he had walked a long distance in a stupor.

The setting for the following stories is the area where I have lived since 1972. To the west is Lake Manitoba which flows via its only drain, the Fairford River, into Lake Winnipeg, passing through the Pineimuta Marsh and Lake before entering Lake St. Martin. Three reservations are located on the western shore of this lake, and from these neighbors I have gleaned a good number of stories over the years.

In the same vein as the previous two stories, a hunter among the reeds sees a huge snake, fires, and runs. Another is on an island, and watches something unusual swim towards him from a distance. He puts a heavy charge into his muzzle-loader, and fires at the creature when it comes within range. There is a violent thrashing in the water of something large, and the hunter hastens home.

Another man, in recent years, told me he toured Lake St. Martin one spring salvaging gill nets that had frozen into the ice over winter. As he was busy working on a net, he spied something strange swimming towards him. It veered away before coming close, but he could see that the head resembled a large snake. About a week later, after he regained his courage to resume the search for nets, he was shocked to see the same as before. Shaken by his experiences, Donald gave up on the lake, and sold his boat and canoe.

A fisherman from the neighboring community of Little Saskatchewan experienced exactly the same thing. As he was lifting a net, something came towards him, raising water plants to the surface of the shallow water as it swam around his net. He saw no head, but the dark body of a 50-foot-long creature rose over a foot above the surface. In fear, the man cut the net, freeing him to take off.

As if one experience weren't enough, he encountered an eight-inch-diameter snake one day on the main road of his community just north of the Band Office. It was partly in the water-filled ditch, and partly on the road. He observed its head to be smaller than the rest of its body, but it had the same shape as a garter—with a similar tongue sticking out. It was also a dark color, between charcoal and green. Rather than wait for it to cross, he turned his truck around.

At about the same spot, a teen on his bike (Richard) also saw a large serpent, but it may have been a different species according to the description of it. The dark snake was also in the watery ditch, and its head was raised about a foot out of the water. It was bigger than a human head, and not diamond-shaped like the garter, but more rounded and flat-topped, like a hamburger bun, with the eyes protruding from the surface.

Two brothers from the same community were paddling their canoe on nearby Pineimuta Lake, which flows into Lake St. Martin. They spotted a stovepipe-sized snake swimming about 100 feet away. Feeling too vulnerable in the open water, they paddled furiously towards some reeds for protection, deeming their .22 rifle virtually useless against a twenty-foot-long serpent. Its shape and green-and-black color reminded them of an oversized garter snake. One brother stated that although most of its body was visible in the calm water, the head remained fully above the water, exposing an orange throat. He estimated the width of the head and body to be eight inches.

Within a couple of miles, two young sisters crossed the lower Fairford River by canoe one evening, and also saw two large snakes. They hit at them with their paddles to keep them at bay, and then ran home terrified.

A man who had been on Lake St. Martin about 20 years ago told me that he saw a snake whose diameter was "about the same as an ice-cream pail." It swam so quickly in its undulating fashion that he was not able to get near it with his 25 H. P. boat.

I understand that snakes are known to propel themselves only with horizontal movements, and yet some witnesses definitely specify a vertical undulation.

On the east side of Lake St. Martin, on a sandy beach that the locals refer to as German's Beach, two Fairford men had spent the night in a tent. In the morning they saw the coffee-can-sized imprint of a serpentine creature that had come out of the lake, circled their tent, and gone back into the water.

From different spots around this lake came a number of reports of large snakes seen coiled up on the shore, with the witnesses giving them a very wide berth.

One elderly lady I had been referred to told of her encounter on the lake shore when she was a young married woman. Accompanied by a child, she had gone berry picking, and noticed what she thought was two piles of tires. Suddenly two long, dark snakes seemed to fly through the air towards the water. The lady and child took off for home, running and stumbling. The next day she had a miscarriage, and lost a baby girl.

This next story has a happier ending, even though the witness might not have thought so. The setting is within a few miles of the previous story—at the north end of Lake St. Martin, and within a few years of the other incident as well. A farmer, recently immigrated from Europe, was putting up hay near the lake, and decided to cool off in the water. Suddenly startled by something that raised its head above the surface near him, he streaked home to the amazement of his wife, leaving his clothes back at the beach.

Whether or not this was a serpent is not definitely known, as creatures other than snakes have been spotted in this lake as well. They are dealt with in the Manipogo chapter.

Folks here relate a story that I found quite shocking at first. A farmer noticed that some of his calves were growing thin, so he decided to investigate. At the edge of the lake he discovered what was sucking milk from some cows—a large snake! About 200 miles to the southwest, beside the Assiniboine River, I heard the same story, told by the granddaughter of the farmer who had seen it happen.

In 2005, two men and a boy were gardening beside the lower Fairford River. They spotted what looked like a huge log that should have floated right by them. As they watched it, they noticed that it was moving across the current towards an inlet where it disappeared from sight. It was described as being about the size of a hydro, or power pole.

This description is used again and again by witnesses in different communities and in different lakes and rivers, so that it is definitely not unique to Lake St. Martin where numerous sightings have been reported over the years. One account was unique in respect to its spiraling motion, but I gather that similar descriptions have come out of Lake Manitoba as well. The size of some of these creatures and their means of locomotion are rather un-serpentlike, although their general appearance is obviously serpentine.

A man from Lake St. Martin was watching three ducks from a window near the lake shore when something like a big snake came up and gulped down two of them.

Also in Lake St. Martin, near the middle of the last century, the head of a large snake had appeared directly in front of a boat filled with people, too close and too suddenly for the driver to see or avoid. The impact even moved the 20-foot yawl out of its path, causing a good deal of screaming and crying on board among the women. Although they circled the area a number of times, guns at the ready, nothing further was seen. The elderly man who recounted this story in 2008 had been present with his father, and remembered the snake's head being about a foot and a half out of the water, revealing a seven-inch diameter, cream-colored throat on an otherwise brown head.

In nearby Lake Manitoba, a woman was sitting in her vehicle waiting for her husband to finish lifting his nets not far from shore. She saw this big "log" go by, against the breeze, so she called to her husband to get off the lake.

On the east side of Lake Winnipeg some men in a boat saw a "hydro pole" which was more than 60 feet long floating on the surface of the lake. They moved towards it, getting rope ready to attach to it in order to tow it home. They soon discovered that it was not a pole at all, but a living creature which quickly disappeared.

Reports of giant snakes that are stretched across trails and roadways are prevalent. Here at Fairford a car full of young people drove down a quiet side road and stopped to move a log out of the way. They soon found that the "log" moved itself off the road. Some years later, on the main highway through Fairford, a woman was on her way home accompanied by her children when they ran over what they realized was a large snake stretched across the highway. Unable to continue driving, the hysterical woman stayed in the car for a while until she regained her composure, and in the meantime, the snake disappeared.

Right near the bridge where Highway 6 crosses the Fairford River, a woman walked along a trail that led to the south shore of the river across from my campground. Suddenly she heard a rustling sound and saw a large snake slithering away into the bush. She ran home terrified, and convinced her husband to move away from the area.

On the eastern shore of Lake St. Martin are some smooth grassy flats. After crossing the narrows of the lake, some hunters on ATVs sped across the meadow through the swamp grass. One of the drivers, Solvie Moar, told me that he suddenly hit something that caused him to be momentarily airborne, but he managed to maintain control of his vehicle. Some time later, on the return trip, he decided to remove the log, but there was nothing there. Whatever it was that his ATV had struck had moved itself out of the area.

At Jackhead, on the western shore of Lake Winnipeg and to the north of Peguis, a man was driving on a new road to visit a friend. As he drove where the road led through a swamp, he had to stop and wait for a long snake to cross. He could not determine its length since its extremities were in the water on both sides of the road. He also told me about a man from the community who shot at a large serpent that was following his boat. That snake was said to have horns of some sort.

In Swan Lake, north of the town of Swan River in west central Manitoba, snakes with horns are a familiar story. Lester Brass told me of his experience out on the lake one spring before goslings were big enough to fly. As he was driving around looking to catch some, a large serpent appeared near the boat. Its head was just under a foot in diameter, and the body was longer than his 22-foot yawl. But what was unique about it were the seven- or eight-inch horns sticking out from the top of the head, making it look like "a jumper buck." It swam very quickly in a lateral fashion. He mentioned that when he was a boy, his brother and grandfather had reported seeing the same thing stick its head out of the river near home.

The other Swan Lake in southern Manitoba also has plenty of giant snake stories, but horns are not a reported feature. One man from there apparently became ill after encountering one as he was crossing the lake one evening.

Snakes chasing after boats is a phenomenon that I have heard from a variety of communities, especially from the bygone days of smaller outboard motors on large boats. Out of Berens River is the account of a snake catching up with a boat and going for the propeller, only to be cut up by it and then disappearing.

I had always wondered why large snakes chased boats, but this story perhaps gives the clue that solves the mystery. It may be that the creature perceives the

STRANGE CREATURES SELDOM SEEN

Swan Lake serpent with antlers
witnessed and illustrated by Lester Brass.

turbulence caused by the propeller as something alive that might be good to eat, or needs further investigation, so it gives chase in order to see what it is and get a closer look at it.

A man from Berens River told me he saw the head of a snake that was about the size and shape of a deer's, and its two horns were about six inches long and six inches apart. The creature submerged when it saw him. Other stories of snakes with horns circulate in this community as well, including ones with a single horn-like feature on the tip of the snout. At Oxford House, a big snake with horns was also seen in the lake.

I might mention here that there are some big water creatures with "horns" reported in various communities, some of which have massive heads and stubby horns resembling those of the giraffe. It may well be that some of these creatures are not snakes at all, but something altogether different. The term that is translated as "snake" in English is loosely used to describe unknown water creatures that bear some resemblance to snakes.

A school principal on the Dakota Plains First Nation near Portage la Prairie was driving home one hot day when she had to stop and wait for a huge snake that was stretched across the road. She told me that it was larger than a stove pipe, and watching it made the hair on her neck stand up. She never saw the head as it was already hidden in the ditch grass. It was a grayish-black color with no pattern. The Assiniboine River, where it likely originated, was about half a mile away.

One summer in the 1980s, three men went ashore on an island in Dawson Bay of Lake Winnipegosis, as they were waiting to go out commercial fishing the next day. They heard something coming down the hill, and saw the tall grass swaying as something came towards them, so they waited to see what it might be. When it came to within 20 feet of them, it stopped at a little clearing where it could see them and they could see it. In the words of the witness, "it looked just like an anaconda or python—like on a TV show."

They saw its eyes, and watched the tongue shooting in and out of its mouth towards them for about a minute, its head about the size of a human's. One man was carrying a long pole that was used to push off their boat, and he dropped it, causing the creature to take off. The terrified men made for the boat, scrambling to be first aboard. Even though it was only late afternoon, they never left the boat again that day, and spent a sleepless night in their bunks, hearing every little noise that the movement of the boat made. They were convinced that the snake had come towards them thinking that they were a source of food.

The witness stated that the experience is still so fresh in his mind that he refuses to go back to that island in summer.

When I visited the northern community of Pukatawagan in the winter of 2008, a lady in her seventies, who still hunts and traps, took me to the spot where she had seen a large snake in the summer of 2006. She and another lady had been canoeing near a rocky shore when they spotted a large snake partly on a flat rock about 20 feet in front of them. When it became aware of them, it slid backwards into the water, and swam away, exposing part of its body. Lena took this opportunity to shoot at it in her fear, and immediately the tail lifted out of the water about three feet. She described the head as being slightly larger than a human head, and its total length as being about 20 feet. Its color was the same shade of light green as the one in the picture eating leaves, but instead of diamond-shaped black spots along its side, there was just a solid black line.

I was surprised to hear of a serpent seen so far into northern Canada, since Pukatawagan is near the 56th parallel, but I suppose that the lakes there also achieve tolerable summer temperatures for their survival, and provide temperate caves for dormant periods.

One summer in the 1990s a man from the Fisher River First Nation, also in the north Interlake, rented a boat from me in order to fish in the Fairford River beside my campground. His brother took some children out for a quick ride first, almost hitting a serpent-like hump in front of them that rose about a foot out of the water, grayish-green in color. Although the one-foot thick body appeared to go down slowly, it created a whirlpool as it disappeared.

This is the first report of a large snake seen in this section of the river in the more than 30 years that I have had the campground. I have always wondered why more sightings haven't been forthcoming, especially with so many fishers gazing upon the waters summer after summer. Obviously, these creatures must not be too plentiful in the river, or don't surface noticeably very often.

The gentleman who shared this story referred me to his sister who had a unique snake story of her own from when she was a girl at Fisher River in the late 1940s. She told me that when she was 11 or 12 years old, she was sent to the store a couple of miles away. To get there, she had to cross the river where it was shallow, and then follow a dirt road. On her way back from the store, she encountered this huge object that was stretched right across the road "about as big around as a wash tub." It had not been moving, but regardless, it struck fear into her, so she slowly retreated and hastened home by a different route.

She remembers its faded alternating black and white stripes. The head and tail were out of sight on either side of the road in the grass and bushes.

A similar report dating back close to 30 years came to me recently as well. Apparently two sisters were driving home to Duck Bay one summer night and had to stop on the highway to allow a culvert-sized serpent to cross. The road is on a narrow peninsula reaching out into Lake Winnipegosis almost 40 miles north of the town of Winnipegosis. Based on other accounts, my guess would be that the creature was simply taking a shortcut overland to reach water again, and therefore was encountered on the highway.

I have also heard numerous reports from hunters or travelers seeing what appeared to be a post or broken tree standing strangely alone, only to find it missing a short while later.

A young farm boy living just outside of Moosehorn had a frightening experience. Now an old man, he told me that he had gone for the cows at milking time one evening, and, instead of going through the usual pasture, he decided to take a shortcut through some woods. He was about to step onto a huge log that lay in his path when it moved, and sent him sprinting away in terror. As I discussed his experience with him, he said he had always assumed that it may have come into the area in a railroad boxcar, as he had not heard of such creatures before. He estimated the snake's length at over 40 feet.

In the same vicinity, and around the same time, a farmer had been driving down a road past a swamp with his small tractor, and had to stop to let a large snake cross over.

A great deal of drainage work has been done in this area in the last half century, and I believe that the relative dryness has kept the snakes from coming inland as frequently in recent years. These large serpents seem to be water-based, and are generally seen in or near lakes, rivers, and swamps. There are dozens of accounts of large snakes seen in that environment, and they are not limited to Manitoba.

In southern Manitoba, around the middle of the last century, a scrap dealer was traveling in his heavy pickup from Winnipeg to Brandon where his business was located. He drove over what appeared to be a huge snake that was lying on the road, but by the time he stopped and turned around, it was off the road and out of sight. Other stories are similar, so it appears that the big serpents are still able to crawl away even after being crushed by vehicles.

A similar incident in southern Manitoba at Sioux Valley west of Brandon took place in the summer of 2006. A man and his wife were going for some firewood near the Assiniboine River when the front wheels of their truck bumped over something in the tall grass. Expecting it to be a log, they were shocked when a huge snake crawled away. Months later, Harold was still loath to show me the scene of the encounter, but the cold fall temperatures convinced him that it should not happen again.

Relatives of Harold, two brothers, also encountered a large snake. They had been in the buffalo compound the following summer when they saw a snake in the grass that was between nine and ten feet long, and close to six inches thick in the middle. The head was considerably smaller, but was "round and flat on top like a hamburger bun, with the eyes near the back." Its back was a dark green, and its sides brown and black, with a shiny, greasy appearance. Man and beast stared at each other for several minutes from a distance of close to 30 feet, the tongue appearing a number of times "but not as often as garters." Jerry, one of the brothers, was not afraid of it, and, knowing that such a rare creature would be a prize, he started towards it, but it made for the water with great speed. He also mentioned that around ten years ago his uncle had been on the water in the same area and had seen a big snake which went under his canoe, almost tipping it.

I had occasion to travel to Alaska in the summer of 2005, and I stopped and made enquiries in every Native community I passed through. On the Sweetgrass First Nation near Battleford, Saskatchewan, I spoke to a brother and sister who separately told me their father's experience of long ago. He had been riding his horse through a ravine across a creek bed when the horse reared and threw him off, all on account of a huge serpent in the way. The sister also told of hearing about three old ladies from their community who had been picking berries in the 1940s. They had stopped to smoke their pipes, and while they were resting, a large snake, bigger than a stovepipe, had come towards them. Also around that time, a lady had walked to the creek to get water, and came across a huge snake. Just before she fainted, she shouted the word "snake" in her language, and told about her experience after she was rescued. On another occasion, ladies out berry picking had encountered a coiled snake that lay about four feet high in the shrubs.

Rumors of large snakes also come out of Alberta, especially in the Saddle Lake and Frog Lake areas.

Matching the picture of a snake reaching up towards a window is a story that is set in the 1960s near St. Laurent, Manitoba. To the east of this town are the Shoal Lakes, and nearby was the farmyard in question. I spoke with Mrs. Fehr who lived there, and this is what she recalled:

One summer day she came back to the house from her garden, and as she was about to go up the steps, she noticed a large, dark snake reaching up towards an open window. She was terrified of snakes at the best of times, and now she had the ultimate horror to deal with. She made it into the house, but the snake remained against the wall. She was desperate to be rid of it, so she called the dog and threw some bread out for it towards the snake, hoping the dog would chase it away. The dog, however, only sniffed at it and left it alone. The snake eventually disappeared, but all the lady could think of was the possibility of having it find a way into her basement, and from there to the main floor. Her extreme fear led to paranoia. Her

In her farmyard just north of Winnipeg, Mrs. Fehr witnessed this unwelcome sight. Artwork by Jarmo Sinisalo.

husband was away working on road construction during the summer, so she informed him that this was the last summer she would spend in that yard, even if it meant living out of their vehicle wherever he worked. She won her case, and soon moved into a new home.

What she did not know was that her family had occasionally seen such snakes behind the barn, but had not bothered her with the information. For example, her son told me he had killed a pop-can sized dark snake with a pitchfork once, and, noticing a bulge in its midsection, squeezed out with his feet over a hundred little ones that were between three and four inches long. When I mentioned that I was investigating snake trails in certain marshlands, he remembered seeing a similar trail on the farm. Years later, as he was driving near Warren, he had tried to avoid hitting what he thought was a four foot stick on the road, but he saw it slither away in his mirror.

His sister told me that as a girl it was her duty to walk to the neighbors now and then to buy eggs. The road led by a swamp, and one day as she was returning with two dozen eggs, walking leisurely and whistling, she looked up and saw a huge snake stretched across the road, blocking her path. The shock of seeing it caused her to drop the eggs, and she was forced to wait until the serpent decided to move along, but she had to throw stones at it to get it going. When she finally made it home, she was told to return to the neighbors and buy eggs again, so off she went, this time without incident.

In the same decade, a slightly different scenario played itself out near the south shore of Cedar Lake, in central Manitoba, where the Chemawawin First Nation (Easterville) is today. Firefighters were working in the area, and a big bulldozer had made a wide fireguard with its angled blade, clearing a double swath down to the dirt. Two firefighters walked along this wide, bare strip of ground, one carrying a shovel, and the other an axe. They came upon a huge snake that not only spanned the whole clearing, but also had its head and tail out of sight in the grass behind the two ridges of dirt and debris. They stood amazed, and pondered what to do. The man with the axe suggested he chop it in two, but the other—who related the story to me—advised against it. So they stood and gazed, until finally the serpent slowly moved out of their way, leaving them with the invaluable memory of what they had seen. Had it moved quickly across their vision, they would not have been able to memorize the distinctive colors and patterns on the creature that you see in the picture.

When I showed this picture to a man from Shamattawa, a remote Native community in northeastern Manitoba, he said that he and his partner had seen the same thing at Sturgeon Lake, in Ontario, east of Sioux Lookout. There had been two snakes, however, about one-and-a-half feet in diameter, and the length of a house.

Snake stretched across fireguard near Easterville, described by Alexander Marsden and painted by Jarmo Sinisalo.

Also on Cedar Lake, some men were returning to Easterville when those watching the front of the boat for floating logs indicated that there was something to avoid in front of them. It turned out to be an eight-inch-diameter serpent that swam away in a snakelike fashion. Another time, some of the same men saw something similar in the moonlight that also moved in the same manner, and its head was bigger than a human's.

Connected to Cedar Lake is another named Moose Lake. These large bodies of water, which are part of the Saskatchewan River system, occupy much of the territory between Grand Rapids and The Pas. It is only to be expected, then, that snake stories from one area would be duplicated in the other.

And there are numerous accounts of four- to six-inch diameter snakes seen here as reported by residents of the Moose Lake First Nation. Some mention seeing large snakes emerging from, or retreating into, holes in the ground. One witness stated that as he approached, a large snake that had its head two feet in the air disappeared backwards into a woodchuck hole. It was a dark grey color.

Another saw a similar snake that was around four inches in diameter, dark grey in color with a white throat. Its head protruded from a beaver hole beside a swamp, and when it turned its head, it "squeaked like a rusty bolt."

Some report seeing large, dark-colored snakes reaching up out of ditches and swamps, no doubt for the purpose of gaining a better view.

What one couple from this community saw did not fit that pattern at all, however. Indeed, not even the color was similar. In Bradley Lake, near Moose Lake, a husband and wife witnessed a six-inch diameter snake reach about five feet out of the water in order to eat the leaves of either a birch or a poplar. On seeing them, it slid back down into the water. As the artist's rendition of the description shows, it was a light green color with "cigarette package"-sized diamond-shaped black spots down the side, about three inches apart. The shape of the head was somewhat similar to that of a garter snake.

Another lady from Moose Lake was picking cranberries beside a swamp one summer when she saw a snake standing up about four feet. It was around three-and-one-half inches in diameter, and was black with small, bright orange spots "that were smaller than your small fingernail."

One man chased a four- or five-inch-diameter snake that stood up about six feet and transformed its head into a wider, flat shape "like the palm of a cupped hand." He hit it with his paddle and fled, leaving the paddle behind. This story is reminiscent of one I heard from a Lake Winnipeg fisherman who had encountered a dark snake on an island. Although it had been much smaller than the one in the previous story, it was described as being very aggressive, and when it tried to attack, it had

This snake was observed by Leonard Nasikapow and his wife Barbara of Moose Lake as it was eating leaves. Painted by Jarmo Sinisalo.

also exhibited that flat, expanded head shape that is so common to the cobra. From the north end of Lake Winnipeg, a fisherman told me he was cleaning his net when a one-and-one-half-inch diameter black snake was about to bite him, but someone grabbed it and killed it just in time. It also was described as having a large hooded head, big fangs, and yellow and green stripes. It was apparently sent away to be examined, and the report had indicated that it was a venomous snake.

In evaluating stories of these cobra-like snakes, it is worth recalling that southern Manitoba is home to the western hognose snake, which hoods like a cobra and pretends aggression when threatened. It is harmless to humans, but is a rear-fanged venomous snake.

The Moose Lake fishermen had an icehouse beside one of their lakes, where blocks of ice covered with sawdust were stored for the summer fishing season. Some men checked on it one day, and noticed a large, long serpent inside, partially covered with sawdust. They just left it alone, but when the time came to begin using the ice, several large eggs were discovered in the sawdust. They were close to ostrich eggs in size, but somewhat transparent "like a plastic milk jug."

Landry Lake is a sizeable lake between The Pas and Moose Lake, and the following incredible account is to have taken place near its shore sometime during the great depression. Apparently two people were in a canoe when a chimney-sized snake wrapped itself around their canoe and crushed it, but they managed to escape to shore. The black serpent had been described as having teeth and horns.

Between Lundar and Eriksdale, around 1941, a farmer led his horse to water where he encountered a large snake. Since he detested the creatures, he quickly tied up the horse and went after the snake with a stick. It disappeared into a rodent's burrow, so the man plugged the opening with a stone. Back at the house, he asked his wife to boil some water, and carried it to the hole accompanied by his wife and son. After he poured the hot water into the hole, the head of the snake emerged about two feet, and then it died.

The farmer took a three tine fork he had brought along, and stuck the middle tine through the head. It was all the two adults could do to pull the long serpent out of the burrow, and, when this had been achieved, the man put the laden fork over his shoulder and dragged the snake to the yard where wood chips were used to make a fire to burn the beast.

There was an obvious bulge in the body of the snake which concentrated in the lower belly as it was carried in an upright position, and, as the fire became intense, this bulge split open, releasing hundreds of two-inch snakes, some of which were still alive. A cattle buyer who happened along watched as the tiny snakes were separated and counted one by one, until the number of them reached 500. He moved on at this juncture, while the count continued to another 500. There were apparently still more, but the official count ended at 1,000. The cattle buyer had estimated the length of the greenish-black snake to be about 12 feet.

The son, now an old man who witnessed the event as a boy, pointed out to me that the swamp on their land had been connected to the Shoal Lakes further south. This was the area where a similar snake had reached up towards an open window. He suspected that the one his Dad killed may have been a "milk" snake, as one particular cow which he always milked often came home bawling in the middle of the day, and at milking time, one teat would never produce the same as the others. He mentioned that he had seen several other snakes about that size over the years as he farmed the land, and had heard reports of marks in the grass that appeared like large motorcycle tracks. He had also heard, like I had, that sometimes it was apparent that whatever had flattened the grass carried a substance that killed it.

The granddaughter of the man from the Lake Manitoba First Nation who frequently fed a Manipogo told of an experience she had heard from her father. He and his father had been hunting near Birch Lake northeast of Ashern when they spotted a trail in the grass that had obviously been made recently by a large snake, since the width of the flattened grass was almost a foot. They had followed it for a while until they realized that they really didn't want to catch up with it.

A man from the Long Plains First Nation near Portage la Prairie, and his son, were hunting beavers in the Assiniboine River in the 1960s. The son, Lawrence, told me that they had put their home-made wooden boat into the river near the Saskatchewan border, and with their five-h.p. outboard, made it back home in about ten days. Early into the trip, somewhere between St. Lazare and the Birdtail First Nation, Lawrence's dad saw something up ahead that he mistook briefly for an upright log—until it moved. He pointed it out to his son, but by this time it had submerged, leaving them to wonder what on earth it was. They decided to use their beaver-tracking skills to shoot it in order to have proof of their sighting, staying behind it about 25 feet. It made such waves at one point that the water sloshed into a beaver tunnel just above the surface, flushing two startled beavers out and into the river. On seeing the creature, however, they hastily exited the water and ran back up the bank.

Only a part of the creature's back became visible again, and it was a brownish-yellow color. The head that the dad had seen three to four feet out of the water, he described as "snake-like and dinosaur-like, and big."

Someone that Lawrence knew had also been hunting beavers in the Souris River, to the south of the Assiniboine. He had been sitting on shore, patiently waiting for the animals to appear, when suddenly the water came up about three feet in front of him, and then went back down. He never saw the cause, but perhaps a sizeable creature had come to the surface and then quickly submerged, causing a mini tsunami.

Some of these stories may be more appropriate in the Manipogo chapter, but not knowing exactly what these creatures are, I will categorize them as snakes for now.

Islands seem to be a popular habitat for snakes. Whether it is islands in Lake Winnipeg or Lake Winnipegosis, or islands in smaller lakes, different sizes and species of snakes are commonly reported by a diverse cross-section of individuals, especially commercial fishers, who frequently stop or camp on them. Often it is noted that there are swamps on these islands where the snakes seem to live, and at certain times of the year, hordes of them can be seen swimming between islands as they migrate to and from their summer grounds. It is here where many large garters have been seen, but some reports have stressed that a variety of species co-habit such places. The head of one was described as resembling that of a Doberman pinscher.

Where I live near the shores of Lakes Manitoba and Lake St. Martin I have heard many accounts of huge snakes seen coiled around small piles of hay (or in recent years round bales) or draped over the top of them, and Lake Winnipegosis has similar stories.

One family described what they saw on the shore at the Lake St. Martin Narrows as a coil of weeping tile that vanished shortly. Beside Lake Manitoba a teen saw what looked like a big tractor tire with smaller tires on top of it. This same description is repeated by individuals with similar experiences beside different lakes. In one case, I heard that men seeing "a pile of tires" on the shore of a small lake in southwestern Manitoba were going to sit down for a rest on top of them, but soon changed their mind.

A man living beside Lake Winnipegosis told me that his uncle, who had trapped years ago in the Duck Mountains, had rested on a log that was partly in the water. Several hours later when he returned to the area he noticed that the "log" was gone, and it appeared that it had been pulled into the water. He realized then that it must have been a large snake that he had unwittingly sat on.

Near where the Dauphin River has its beginning at the north end of Lake St. Martin, about half a century ago a large group of Natives from the Lake St. Martin First Nation made their way downstream by canoes to engage in their typical summer hunting, fishing, and Seneca root-digging activities. Some stopped at an isolated farm on the east shore of the river in order to exchange some goods, as was the custom. The young wife gathered that the trip was to be extensive, so she was surprised when one party dropped in a few hours later to report that the expedition had been aborted on account of a large snake. A few miles downriver from the farm was a large boulder on the shore, and here it was that they reported the killing of a serpent with their spades. Knowing the landmark well, Pauline decided to check out the story. She jumped on her pony and found things just as she had been told. The tail of the snake was still twitching, and her horse was reluctant to draw close to it. The 11- to 12-foot-long serpent was indeed hacked up quite badly, but the color remained vivid in her mind. The back of it was dark, but the underbelly was a light brown—like a pinkish beige.

An outstanding feature of this snake, she told me, was that it had a round, flat head "like a hamburger bun" with the eyes on top. (This description seems to match the one given by the teen from nearby Little Saskatchewan—of a rounded, flat-topped head with eyes protruding from the surface.)

Speaking of large garters, there are numerous reports of pop can-sized ones. One individual, as a teen, chased a six-footer on Peonan Point at the north end of Lake Manitoba, and couldn't keep up with it. Another farm boy saw a similar one on the east shore of the same lake, and said, "it moved like lightning."

A highway parallels the Dauphin River on the way to Lake Winnipeg east of Gypsumville. It has been there for less than half a century, but there have been a number of motorists who have encountered large snakes on the road.

Before I change from the Lake St. Martin stories, I must mention some of the experiences that swimmers have in that lake. The most common reports are of something unusual and unidentifiable lurking a safe distance away. Sometimes it is *not* a safe distance away, and children are then quickly shooed out of the water. Occasionally some live creature is felt touching feet or legs below the surface. One lady recalled her experience as a girl at the beach on the north shore of Lake St. Martin in the 1970s, saying that she stepped on something stovepipe-sized with one foot, and again with the other. Feeling it to be something alive, she exited the lake, pronto.

Two teenaged girls were on a beach on the eastern shore of Lake Winnipeg in the summer of 2007 and had the same experience. They discovered this "log" lying on the bottom, so they stood on it, enjoying the feel of a "scaly log with a loose jacket." When they went back to it after calling others over to join them, it was gone, but ripples in the water indicated something moving away. One mother, who was photographing the event, told me that she saw the girls' expression change drastically as they came to the realization that they had been standing on a large serpent.

Also on the east side of Lake Winnipeg, two boys had gone swimming in an area where they had been discouraged from going, since large snakes had been seen there in the past. One boy had jumped in and was treading water as he was waiting for his partner to follow suit. In the meantime, something long and cold passed by his legs. His father told me that he had immediately gone home, leaving his shoes behind, never to return.

Beaches on the shores of Lake Manitoba have such stories as well, but fortunately for swimmers of all sizes, there don't seem to be any reports of attacks. Therefore, these stories may not be harbingers of something to fear so much as relating for our awareness the existence of benign creatures that share our waters. There is no doubt, however, that encounters of this nature can raise serious concerns, but fortunately the record so far indicates that they are needless concerns.

A common observation seems to be that whatever lives in the water is attracted to human activities like swimming and splashing, causing generations of elders to discourage their children and grandchildren from swimming in the lakes and rivers. If we examine Marcel's experience when he was washing his hair in Lake Manitoba, and the streaker's similar activity in Lake St. Martin, then we can agree that some aquatic animals are lured by the very nature of such activities. That is why I feel I achieved almost instant success when I used a simulation technique to get their attention.

I have heard of several farmers who chased after large snakes with their horse-drawn mowers. One man from St. Ambroise was cutting hay near the lake when a stovepipe-sized snake stood up about four feet high in the grass. He forced his reluctant horses after it, but it got away. Another farmer was cutting in a swamp in the Ashern area when he spotted a six-foot-long black snake which escaped into a sink hole.

Near Amaranth, on the west shore of Lake Manitoba, a farmer long ago was using a horse-drawn dump rake. An eight-inch-diameter snake got wrapped up in the coil of hay, causing the horses to bolt. He managed to release the load and avoid disaster.

On the east shore of Lake Winnipeg, years ago, a large serpent apparently got between a young child and its father. He managed to kill it with an ice chisel, but it is not known whether or not it showed signs of aggression.

Although the existence of large snakes in Canada is not compatible with prevailing scientific knowledge, the sheer number and variety of reports is overwhelming evidence to the contrary. Just as the known small snakes survive the cold, so do the larger ones. Most people who consider the question think that cracks and caves in limestone which are accessible from underwater are their likely hibernacula.

Then there is the question of what they feed on. Most of their food must come from in the water, as they are not often seen on land. When they are spotted out of the water, they are often seen resting and sunning themselves. Fish are quite possibly one of their staples, besides a host of small water-dwelling creatures.

A snake story that really took me by surprise the first time I heard it was in regards to the flying or jumping snakes. A husband and wife, living at Moose Lake, both had seen the same kind of snake, one at Moose Lake, and the other at Grand

Rapids. In appearance they were similar to garters, but instead of tails that tapered to a point, they were blunt "like a thumb." They appeared to fly "like an arrow" for 10 to 15 feet, launched by a tail that was anchored against the ground. Perhaps the "arrow snake" that is mentioned in Isaiah 34:15 in some versions is this type.

There is some knowledge of "flying snakes" in the Lake Winnipegosis region as well. One man from the Pine River First Nation on the west shore of the lake told me he saw a blunt-tailed snake that jumped seven to eight feet. These snakes have also been heard of on the east shore of Lake Winnipeg, and in addition to the characteristic of flying, some are purported to have legs. However, while the "flying" or "jumping" snakes are spoken about here and there, ones with legs are rarely mentioned, and perhaps I should not even be doing so here.

For lack of corroboration, I was not even going to mention the report of a thin flying serpent that actually had a pair of wings, but since I have heard that there are similar reports elsewhere, I will include it. A gentleman living on the east shore of Lake Winnipeg claims he saw such a thing fly over his yard, but he could not find where it landed. From the west side of Cedar Lake comes a report of dark, white-bellied snakes that will jump from up in shrubs to the ground and "turn around in the air like you throw a stick."

A farmer's wife from near Fork River, just south of Winnipegosis, described a snake that she saw in the early 1950s when flooding revealed more denizens of the lakes than usual. It was about three feet long with a three-inch diameter head, and was brown yielding to gold in the sun. What was unique about it was that it had antennae-like stumps on its head, and a forked tail! She said the antennae were about one-quarter of an inch thick and one-and-a-half inches long, and the tips of the forked tail were about two-and-a-half inches apart and curved upwards to at least three inches above the ground. Her father-in-law had seen the trail of a very big snake in sandy soil the same year.

Another type of snake that seems to be quite different from the ones we might expect to find in Canada is one which, instead of slithering along in the usual manner, moves with a sideways motion, much like the sidewinders of the southern deserts. One such snake, red in coloration, was seen sidewinding on a road by several children at Sioux Valley.

Another report indicated that someone walking barefoot on an isolated island beach in Lake Winnipeg received a serious inflammation when a small burrowing snake-like creature came into contact with the bare skin. A doctor had commented that a timely evacuation by plane had quite likely spared the individual's life.

Mention has been made of snakes seen using holes dug by woodchucks and beavers. Now that I have heard about the tunnels made by the giant beavers, I wonder if the giant snakes use them. Many of the Native people in southern Manitoba who know about the large tunnels insist that they are inhabited by large serpents. This information has been handed down through the generations, and may very well be correct, but what is not so well known concerns the creators of the tunnels. In the north, however, the tunnel-making habits of the giant beaver are understood.

Most communities that are situated beside large bodies of water have stories of boats hitting unseen objects. A good number of people have told me personally that their boat—or more exactly, their motor—hit something big under the surface, sometimes making it pop up, and in a few cases ripping it right off the boat. Each report stressed that it was not a hard object like a log, but rather something softer, like a creature of some kind. Large serpents were generally suspect.

At Oxford House a boat had flipped completely after striking a submarine object, but large water creatures other than snakes are also seen there.

I have alluded to snake tracks briefly, but I should include a recent phenomenon from the area where I live. In the summer of 2007 I was told about some unusual tracks in the grass on Peonan Point, the peninsula at the north end of Lake Manitoba. Some Native hunters had seen them and found them significant, especially because there were so many of them. When I checked them out, I was surprised by what I saw, as there were a great many paths in the grass. Part of the meadow had been hayed by a farmer, so it was obvious that the trails had been made before the grass had been cut, since the mowing, raking, and baling had not disturbed the flattened grass at all. It actually looked as if a big motorcycle had driven back and forth, flattening the grass so much that it remained so over winter. All the trails went in an east-west direction, with a poplar bush being to the east of the narrow meadow, and Lake Manitoba to the west almost half a mile away.

I contacted a well-known trapper/hunter who had worked the area many years ago when he was hired to deal with bears and wolves that were harming farm animals. Rick Lindell brought along another experienced hunter, and together we examined the meadow. There were no four-footed animal footprints to be seen on these narrow trails, so the possibility that these were just ordinary game trails was ruled out. The men admitted that they had never seen or even heard of such trails in their lives.

I recorded the direction of all the pathways in one part of the meadow, and the direction of the flattened grass indicated that half of them were made by a smooth, heavy body moving in an easterly direction, and half westerly, back towards the lake.

In the summer of 2008 we monitored the same area again, and were pleased to see tracks as before. Flooding due to extreme rainfalls made the tracks less distinct than the year before, and fewer in number, but there was no doubt that something serpentine had crossed the same meadow as the summer before.

An impressive "snake trail" crossing through bush near Lake Manitoba.

The packed "snake trail" that remained even after haying was done.

On July 7, 2009, five ATV enthusiasts who were exploring the area joined my grandson and me in checking for snake tracks on Peonan Point once again. We found a good one that clearly matched those from previous years, so it was obvious to all that only a large serpent could have made it. Two days later, when the visitors were out on their own, they encountered two tracks that were large and plain to see in a marshy area near the east shore of Lake Manitoba close to where I live. The tracks had crossed each other occasionally, and the grass was bent flat in the direction of the lake, indicating that a pair of large snakes had returned to the water from wherever they had been.

I have been able to monitor the tracks every year since then, but the first summer was the only one that yielded a large number—possibly several dozen. The following seasons produced fewer that were clear enough to be considered snake trails, leaving us with many questions that may never be answered fully, especially since these unique migrations are so difficult to monitor.

I understand from speaking with a few snake-keepers and experts that there is no apparent need for large, tropical snakes to leave the water, as shedding, mating, and eating can all be water activities. These Manitoba snakes, however, may not be like the known tropical snakes, and as such may have a reason to be on land for a short period of time each summer. There seems to be no doubt that these tracks match the ones spoken about by a variety of witnesses over the years who also concluded that they were made by large serpents.

In the early 1960s, a man had been out with his family digging Seneca root near Waterhen where Lakes Manitoba and Winnipegosis join. He had crossed a swamp to reach higher ground where he spent some time in digging, and needed to return by way of the same swamp. This time, however, he encountered a fresh, wide track of flattened grass that had been made since his earlier crossing perhaps a few hours before. His son, Leo, remembers his dad saying that the hair on his neck had stood up as he realized the size of the creature he had just missed. (I can identify with this gentleman, for as I walked through the tall grass checking the strange trails on my own one day, I had the urge to take my leave of the place as my imagination created scenarios of what might happen if I suddenly stumbled upon one or two of the creatures.) The distance between his hands indicated a trail that was well over a foot wide, a measure that could conceivably translate into a body whose diameter could be close to two feet. This could be perceived to be about the size of a small culvert, which was the description used by the two ladies who watched a serpent cross the road in front of them on the opposite side of Lake Winnipegosis.

An additional note regarding the track in the grass indicated that the grass was "dead." This is an observation that I had heard from others, but which I did not notice on the smaller tracks I have seen. It may well be that this much larger serpent, if it indeed is one, which seems to be in a league all its own, has a substance on its skin that causes the grass to discolor almost immediately after its passage. It is also quite possible that the serpent "the size of a hydro pole" that Leo saw holding its head above the reeds near the shore of Waterhen Lake about 30 years ago was the same kind. The head, which was similar in shape to that of a typical garter, reminded him somewhat of an elk's head, only with no ears, and a dark, shiny appearance. It had disappeared during the time that he turned and reached for his gun.

If April of 2006 was a memorable time for me because of my sighting a giant beaver, then the summer of 2008 was a highlight for the sighting of a giant snake. I had built a cabin and observation deck on an old pontoon boat a number of years ago, but I had never found the right opportunity to use it for its intended purpose, which was to attract and photograph giant snakes in nearby Lake St. Martin. In August I was finally ready, with a crew of five—my son Myron, grandson Tyrell, a friend Gordon, his son Chad, and I.

It was a calm, sunny day, ideal for being on the water. We launched at the Lower Fairford River, barely a mile from the lake. Near Big Fisher Island, we decided to anchor and set up the equipment that I hoped would attract the creatures to us. From the numerous stories of strange objects watching swimmers from a distance, I realized that a device simulating swimming and splashing might lure the beasts close enough to be recognizable. Sure enough, within 20 minutes of starting up the unit, Gordon spotted something long breaking the surface about 75 yards to the south of us where he was facing. It was not impressive at that distance, barely showing above the glassy surface, but it was obviously a living creature, about 15 feet of it, making a visible wake as it moved easterly with no apparent sign of locomotion. In a couple of minutes it disappeared, and in the excitement of actually seeing what we had come for, my camera had been forgotten in favor of the binoculars.

We waited for another 15 to 20 minutes before we saw it surface again, this time coming directly towards us, but again staying back about 75 yards. It submerged again within the space of several minutes, and showed itself for the third and final time about half an hour later in the opposite direction, on the north side of us. It had either swum under us, or, more likely, around us, and was now heading in a westerly direction, giving us the benefit of a good broadside view with the sun behind us, shining towards the long dark shape. Again the binoculars took precedence over the camera, but being only a small digital, I was confident that we could repeat this experience when better camera equipment became available.

Unfortunately, a variety of circumstances prevented further research that summer, so 2009 became the target year when perhaps proof of the existence of giant snakes might be secured. Although several trips were made on the lake in the summers of 2009 and 2010, the weather was seldom ideal for snake watching.

A good number of people from the area called or dropped in to discuss our sighting in Lake St. Martin. All of them had either witnessed something like we did, or heard stories of the giant snakes.

One lady from the Lake St. Martin First Nation wanted to share her experiences, so she told me of her two unusual sightings. The first one occurred when she was a girl walking along a path that had high grass on either side. At eye level she had seen a snake "fly" across the path, but she hadn't told anyone about it. Then, in 1997, when she was walking into the school where she taught, the sound of a spinning rope nearby caught her attention, and there, about five feet away, a garter snake was standing upright, about two feet tall, its midsection spinning in a circle like one would spin a rope, making a whipping sound. It startled her so badly that she ran out of the school screaming. It closed for good shortly after that, as it had had more than its share of snakes gaining entrance year after year.

The 43-foot-long snake fossil that was recently unearthed in a Colombian coal mine may very well be touted as the mother of all snakes from the distant past, but even some reports I have mentioned would suggest snakes that size are very likely still present—in unlikely Canada! (Not to mention one that required three caterpillars to push its carcass into a river in Latin America, where a friend's father-in-law had lived.)

Serpents are a prominent subject in pictographs and petroforms in the Canadian Shield especially, as well as in certain areas to the south of it, so it may be that the inhabitants of this land have always been cognizant of the existence of large serpents the likes of which we hear to this day.

I had always hoped that some early explorers or traders had seen or heard of these giant snakes and had recorded the information in their journals. In the summer of 2009 I was finally directed to a book on the memoirs of Louis Goulet, entitled "Vanishing Spaces." A Metis guide from southern Manitoba, Goulet was traveling in the area of the Missouri River in Montana when he had two fascinating encounters in the summer of 1883. In the first one, his horse spooked and almost threw him off when a big snake crossed a trail in the woods right in front of him. He didn't describe its size, but called it a monster when he said that he couldn't get a shot at it in time. When his skeptical partner wanted proof of his allegations the next day, they returned to the site. Louis refused to follow its track into the woods, so the partner went alone, returning a short time later with the report that the snake was lying in a heap where he was inclined to leave it alone rather than risk trying to shoot it.

The second encounter was somewhere "past Fort Benton" where the two men hoped to shoot one of five mountain sheep they saw. However, they had all escaped except for one which was hobbled by a large snake that had wound around its body, its front foot in the snake's mouth. After they shot the serpent, but before they had a chance to stretch it out and measure it, some Native men came along and took it away in order to skin it. Louis estimated the diameter of the body to be a foot, and the length between 12 and 16 feet.

I spoke with a lady from Page, Arizona, on a recent trip, and she mentioned seeing an eight- to nine-inch diameter snake that caused a herd of cattle to part during a roundup about 20 years ago. It was both gold and black, and had not been near any water even though the waters of the Colorado River systems were not all that far away.

In the summer of 2010 I heard rumors of unusual snakes seen in a ditch adjoining the swollen Fairford River right beside the Lower Fairford Bridge—an area where numerous big snake sightings have taken place over the years. Upon investigating the incident, I found that the story greatly exceeded my expectations, not only in the uniqueness of the sighting, but also in the ability of the witness to sketch what he saw.

Michael Sumner had taken a walk, which he often did, to Lower Fairford where the short river which begins near my campground beside Highway Six flowed under the bridge before entering Lake St. Martin. Walking a few steps past the bridge, he heard a splashing in the ditch, and saw not one, but two similar-sized serpents intertwined in the water. They were most unusual in that their heads were more elongated than the average serpent, giving them a dragon-like or crocodilian appearance. Furthermore, they both had little "antlers" protruding from their foreheads, about two inches in length, and the size of a little finger. As to length, Michael estimated they were between eight and nine feet long, and their maximum girth was at least five inches. They were black with rows of white dashes along the bodies, and their forked tongues were reddish-orange.

Michael stopped a car that happened by, and invited the young lady to share the spectacle with him, but when she saw the object of his attention, she drove off in a hurry. After watching the snakes for about ten minutes as they moved along the ditch towards the river, Michael noted that when they entered the current of the river, they disentangled themselves and swam separately along the shore. It is generally assumed that the pair was using the shallow quiet waters of the ditch for the purpose of mating.

Last, and definitely not least, is an incredible story that I heard just recently involving a serpent of hitherto unheard-of proportions in our country. In the Cumberland House area of eastern Saskatchewan, the story is told of a lone hunter

Unusual snakes beside the Fairford River, seen and sketched by Michael Sumner.

who was out in his canoe near a shore. He heard the sounds of something large coming towards the water, and saw trees and shrubs moving in its path. Hoping to see a moose's head emerge from the foliage, he trained his rifle on the spot where he expected it to appear—but instead, the massive head of a huge serpent appeared near the ground, its tongue flicking in and out, and red flaps above its eyes adding to the scary sight. Realizing that his survival was perhaps dependent on his immediate actions, he instinctively lowered his gun and fired, hitting the snake in the head. It writhed and thrashed, exposing its white underbelly—but the hunter hastened away.

He knew where others were camped a few miles away, and joined them for the night—keeping them awake as he maintained a large fire in case the monster followed him. In the morning all went to the site to see for themselves—and sure enough, the area was flattened and bloodied, corroborating the story that a barrel-sized grey serpent with black diamond patterns was wounded on that spot the day before.

IT IS INCOMPREHENSIBLE that there should be such a dearth of information on alleged giant snakes in North America when there is such multiplicity of sightings and stories about them.

Although there is no way that I can vouch for the truth of any of these snake stories other than my own, the fact that most of them are not isolated reports gives me some satisfaction. It is always a delight to hear a story that matches or corroborates previous ones, for as those numbers keep increasing, the overwhelming evidence tends to speak for itself.

It is high time that we look beyond conventional wisdom that is so limited in its scope, and embrace the experiential wisdom of witnesses. When conventional wisdom says that the garter is Manitoba's largest snake, but hundreds of eyewitnesses disagree, what are we to believe?

THE SASQUATCH

At first I resisted the idea of including the Sasquatch, also commonly known as Bigfoot, in a book which addressed all the unusual creatures that had been brought to my attention. I reasoned that since there were plenty of books about them already, there was no need to belabor the topic. However, as I began traveling extensively to all parts of Manitoba, I was given innumerable accounts of these creatures by folks who were convinced of what they saw, and wanted to share their experiences with me. I became increasingly impressed with the consistency of the reports, and the unique and often dramatic nature of the encounters. So, since the creatures met the criteria for the scope of this book, I decided to glean the best of the reports, and include them in a chapter as well. I realize now, however, based on the accounts I have noted, and new sightings that surface each year, that an entire book devoted to them could not contain all the stories. As a result of hearing so many intimate accounts by numerous eyewitnesses, I must say that I have become somewhat attached to this species, whose intelligence far surpasses that of any of the other little-known creatures. Not only is its body so similar to ours in many respects, but also its interestingly unique behaviors are varied and unpredictable like ours.

My first recollection of a Sasquatch sighting took place virtually in my back yard in 1979, about seven years after we moved to Fairford. Some Native people from a neighboring reservation claimed to have seen a huge, hairy creature, and enlisted the help of the local Natural Resources office in an effort to surround the beast. It eluded the heavily armed posse who thought they had it contained in a certain area, and even the use of an airplane did not result in any further evidence of it. A local farmer, whose own experience you will hear about later, followed its tracks through a hayfield, and found an impression where the massive frame had rested against some round bales. I recall being only mildly interested in the story at the time, but we as a family did go to look for the footprint that was purported to be visible in a ditch in a certain place—which we never found.

I did, some years later, speak with some of the witnesses, and their stories did seem quite convincing. I discovered also that the Conservation Officers had taken a plaster cast of the footprint, and I eventually got permission to get another one made from it. It was not an unusually large print, measuring between 14 and 15 inches in length, but it was huge in comparison to a human footprint. I lost the replica in a house fire in 2002, but the original is probably still with the owner.

As a result of their exposure to incidents such as I have just described, I find Natural Resources personnel less skeptical about the existence of the Sasquatch (and other unusual creatures) than many other folks whose judgment stems more from personal opinion than experience.

I became increasingly intrigued by rumors of sightings here and there, but it was the persistent influence of the farmer from St. Martin that helped shape my impressions and perspectives. And persistent Paul Shabaga was. Just why, no one knew for the longest time. As he frequented the coffee shops after his retirement, his focus was mainly on Sasquatch stories, and he did not take kindly to any opposing opinions on the topic. Why an intelligent, successful, and prosperous farmer would be so obsessed with such stories was a mystery for years, until finally he threw caution to the wind and decided, in his old age, to come clean.

He began sharing his experience with those willing to listen, and he never tired of telling his story, even to total strangers.

Apparently, in 1941, at the age of 17, he had gone moose hunting in the Basket Creek area west of Gypsumville with a couple of older men. It had been in November, and some snow was on the ground. He hunted one area by himself, and wounded a moose which he tracked for some time. He spotted a brown form behind some willows, and taking it to be the rear end of his moose, he decided to take a shot. It went down, and when he arrived at the spot a few minutes later, he was shocked to find a big, hair-covered, ape-like creature lying before him.

He had already learned not to trust animals that appeared to be dead, so he kicked at one of its hands to check for a reaction—and there was none. What he had thought to be the hind end of a moose must have been a Sasquatch standing with its back to him, its neckless head bent down looking at the bloody moose track.

Paul told me that he had no idea what kind of animal it was. He had heard his dad make mention of wild men that supposedly lived in the Caucasus Mountains of Asia which were near the homeland of his forefathers, but this creature did not match the picture that he had in his mind. In fact, he said that it would be 35 years before he would really understand what it was that he had shot.

He had left the scene and never returned. He told no one about the incident, not even his hunting partners or his parents. He had mentioned it to an older brother

Artist Jarmo Sinisalo's well-accepted illustration of a Sasquatch shot and described by Paul Shabaga.

when he returned from the war, and the reaction he got caused him to keep the story to himself. What with its human-like appearance, and the fact that it was during the war, Paul said he was not inclined to reveal his experience. Not, that is, until he heard other reports of Sasquatch, and came to realize what it was he had killed. His personal, unique experience explained his obsession with the topic in his later years as his pent-up knowledge finally spilled out.

Paul was a frequent visitor in my campground, and he would keep me informed about the latest sightings and rumors that he picked up. Year after year he would tell and retell the stories—which was excellent for me. Annually I would throw out a question about his own experience to see if I could trip him up in some way, but there was never any variation or hesitation in his account. It was always so detailed and consistent that I have no doubt in my mind that he shot and killed a Sasquatch in 1941.

And Paul is probably not the only one who ever shot one in Manitoba, but because he was willing to endure the consequences of the telling, we now have his story with us.

He was aware of other local rumors that pointed to similar experiences. He suspected that an acquaintance who intimated that he had shot at something unusual may in fact have shot one as well. Paul had also heard of a hunter from Fairford who had packed up his family in a hurry after shooting at a creature that had followed him in the woods.

Another rumor that both Paul and another individual shared with me also originated in the northwest Interlake region of Manitoba—that a trapper using an illegal substance to poison animals subsequently found two huge, ape-like carcasses lying close to each other near the bait.

In Nelson House folks tell of something their ancestors experienced not that long ago as they were traveling on foot. They encountered what is presumed to be a Sasquatch, and had to kill it with an axe in self-defence. Another incident involved a man who shot a Sasquatch and then buried it because it had been bothering him.

On many occasions hunters apparently found the creatures to be too aggressive for their liking, so they killed or wounded them. A good number were shot at through the walls of tents or cabins when they refused to go away. Sometimes the creatures banged on cabins to let the occupants know that their presence was not appreciated.

Often hunters or trappers had logs or stones hurled towards them, and sometimes these actions were accompanied by unearthly screams and foul smells. Such actions by the elusive creature would often continue until the people were well on their way home and out of its territory.

The consensus seems to be that Sasquatch don't normally try to hurt people, but generally want them to clear out of their domain. That is why they are seldom known to close in on their quarry, but stay back a certain distance—usually out of sight—and just scare the intruders away.

A rather unusual encounter, which I heard about from some Native friends (and from Paul) originates somewhere to the east of Lake Winnipeg. Hunters apparently heard a loud commotion in the bush, and found a moose doing battle with a Sasquatch. The hunters shot the two-legged creature, and then buried it with stones.

Another story that took place in northeastern Manitoba was told to me by the owner of Healey's Lodge at God's Lake Narrows. A priest, whom she had known, told her that he had been traveling with some Natives by canoe when they had come upon a female Sasquatch carrying her baby. The hunters had wounded her, but she had escaped, making crying sounds as she went.

Mrs. Healey's two sons had their own story to share. They told me that they had gone to their remote outcamp for moose hunting one fall, and had pulled their boat up on shore, still loaded with moose meat, rifles, and containers of gasoline. They had secured the anchor well, and left everything in place for the night. In the morning the boat was missing, with no sign of it anywhere. They checked far down the shoreline just in case the boat had somehow drifted away. As they returned, they noticed ravens circling an area in the bush on the opposite side of the river. They crossed over, and immediately found evidence of something having been dragged through the shrubs near shore. About 100 yards from the water, they found their boat, still containing the meat, guns, and supplies, with only several gas cans missing.

They spent the better part of a day getting everything back to the river, all the while shouting obscenities at the creatures who refused to come back and help them with their work. There was no doubt in their minds as to who, in such an isolated area, would have the desire or strength to pull the heavy boat across the river and up the bank in the middle of the night. They found the unopened, unpunctured gas containers half a mile away the following spring.

The brothers also recounted the experience of some trappers they knew who had shot a cow moose one fall at the trapline. Since they needed to return home before they could attend to the meat, they covered it well by putting logs on top to make it less accessible for wild animals. A jacket with its human scent had been hung up there also, as an additional deterrent, but when they returned a few days later, the logs were thrown aside and the meat—and jacket—were gone, with only big tracks left behind to identify the thief. Native guides will apparently refuse to spend the night in the area, and need to be flown home instead.

Ironically, the creatures have also been seen right in the community of God's Lake Narrows. One lady told me that she had driven her truck near the school very early one summer morning, and saw, within a stone's throw, a seven to eight foot tall, dark-colored hairy creature that walked with a swinging gait. It had a flat nose and a cone-shaped head, and on seeing it, she had felt the hair rise on the back of her neck. She hadn't gone camping for two summers after that, or driven the truck alone after dark.

As I was waiting to board the train for Pukatawagan in the summer of 2008, I met another lady who had grown up at God's Lake Narrows, and she also had an encounter with a Sasquatch—and a most dramatic one at that. Marlene had grown up living with her grandparents, and as a five-year-old had gone to the fall fishing camp with them as usual. This time, however, her new baby sister was with them, spending most of every day in a hammock between two trees near their tent.

Grandpa had been away fishing on this particular day, and Marlene was watching her granny doing something with fish down near the shore. It was her task to check on her sister periodically, so when she was told it was time to do it again, she took the path up the slope to their campsite. When she raised her eyes upon reaching the hammock, there, across from her, stood a huge hairy creature on two legs staring down at her. Marlene remembers its long eyelashes, blinking now and then as it gazed at her, and then down at the baby, and then back at her—back and forth, its noisy breathing being the only sound it made through an open mouth that revealed big, brown-stained teeth.

Its only movement was a shuffling of its feet and a slight shift of its head as it alternated its gaze, so there was nothing menacing or fearsome at all in its manner. Marlene was therefore not afraid of it, but told me she had felt an overwhelming awe considering its imposing size and strange appearance. She marveled, however, that it made no attempt to take the baby. When it dropped its head more than usual after what seemed like several minutes, she felt it was time to bolt and notify her granny about this strange phenomenon. When she turned back for a final glance after a few steps, she saw that it too was turning to leave.

Granny was quick to come and check out the story, but by the time she reached the hammock, there was nothing to see but the sleeping child. Her comment was that "Keegohogee had been there, and it had not harmed their baby."

When Marlene's grandparents would try to scare her now and then using the creature's name, saying things like "come inside now or Keegohogee will get you," she was never intimidated by it since she harbored no fearful feelings towards the creature. Although she watched for it in the years following her encounter, she never did see it again.

While visiting the community of Cormorant, northeast of The Pas, I was referred to a man and his younger sister who had spent some time as children living with their grandmother at Bird, northeast of Gillam, on the Churchill rail line. The two of them, and their baby brother, had been playing outside the cabin when a huge form, taller than the outhouse, had walked towards them. It advanced to within a few steps of the girl, who stood mesmerized by the sight, until her mother drew her into the safety of the cabin. They quickly covered all the windows and waited in fear, hoping the creature would go away. Instead, however, it began pounding on the wash tub that was upside down on the roof of the porch. Eventually everything was silent, but the family waited a long while before they ventured out of the house.

Some days later, after the grandmother had gone to lift a net, she came running back, thoroughly drenched, and reported that the creature had come after her. The siblings recalled her saying that its appearance was pretty much an annual occurrence.

Marlene meets a Sasquatch. Drawn by artist Dennis Sinclair, for the author's upcoming children's book, *The Girl Who Met a Sasquatch*.

Her running through water instead of around it reminds me of a similar episode that took place much more recently in southern Manitoba. I was visiting the Birdtail Sioux First Nation on a Monday afternoon in 2005 for the first time when I learned of an incident involving a Sasquatch from the day before. I tracked down the 17-year-old witness, and he told me this story:

Donovan had gone deer hunting on Sunday afternoon across the broad Assiniboine River flat, leaving two younger boys behind at the edge of the valley. When he reached the river on the far side, where numerous deer were grazing, some of them suddenly began running in his direction, so terrified that they almost ran him over. He wondered what the cause might be, so he kept walking to find out. Soon he heard something moving in the woods across the river, and then a large dark form appeared. It walked up to the river, and put its long arms around two trees that stood about six feet apart. As it stared at him, it pulled the two sizeable trees together with its powerful arms, and then released them. Donovan was watching through the scope of his borrowed .22, and was impressed by the muscular build of the creature. He was certain it was ten feet tall and had an upper body width of four feet.

But after it shook its body, and the trees with it, it put its head back and let out a powerful and fearsome scream that lasted about half a minute. That is when Donovan said his eyes began watering uncontrollably, and he fled the scene. Even though there was a flooding, icy river safely separating him from the beast, he splashed and swam through deep puddles in his haste to get back across the valley where his friends awaited him.

He paused once to look back after he heard the sound of a stick pounding on a tree, and saw the huge creature moving up the high bank among the trees. When he reached the other boys, they asked him why he was crying, and he insisted that he wasn't, but that it was what he saw that made his eyes water. He pointed across the valley, and there, at the top of the far bank, stood the creature outlined against the sky.

We arranged to check out the area across the river the next day, so I worked on rounding up a boat, trailer, and ATV. Early the next morning we set out for the river, the reluctant youth, his grandfather, and I. The last two days had been very warm, so the snow that we hoped would contain evidence was almost gone. Neither did we find any hair where the creature had embraced the trees. The "awful, sewery, fishy" odor of two days before had dissipated as well.

They told of a trapper who reported seeing deer skulls placed up in trees in the area where we searched, and also mentioned a pile of bones nearer the community that saddle horses had spooked at during a trail ride. Numerous dogs had disappeared from the community one year, and some of their bones were thought to be on that pile.

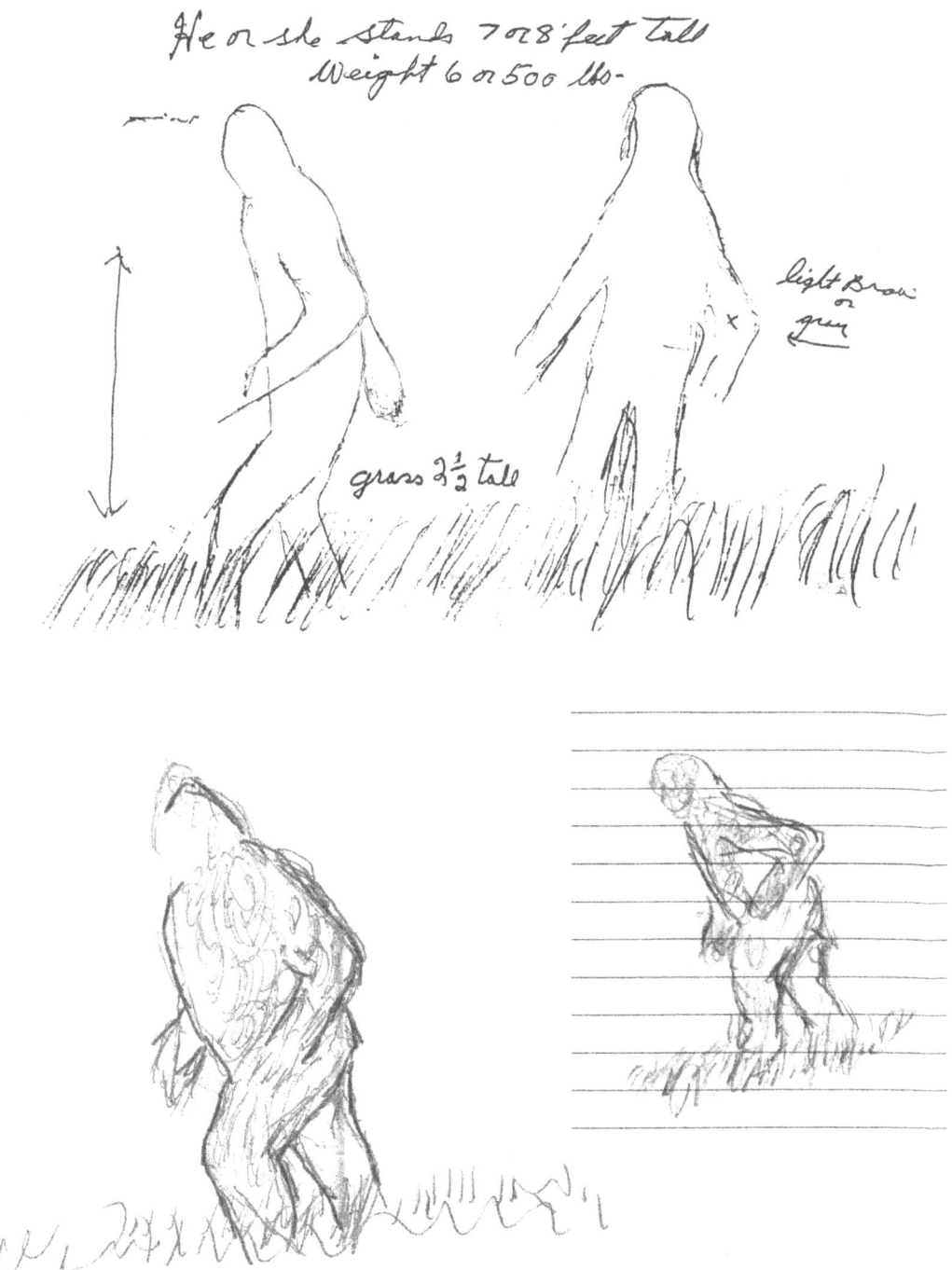

Gordon Stagg's initial and reconsidered Sasquatch sketches.

One young lady in that community told me that she had been alone in her parents' house with her son when something banged on the upper walls of the two-story building after dark. The pounding had taken place on three sides of the house in a very short time, indicating something tall that moved around quickly. She phoned her mother immediately, and as they were talking, she went outside where an awful odor met her, and then a terrible scream split the air, so loud that her mother could hear it over the phone. The next morning they saw a huge footprint in the mud near the steps.

In 1996, on October 8th, Gordon Stagg Sr. and his wife Genevieve left home early in the morning to see if they could bag a deer near the community of Fairford where they live. As they drove by a pasture on the reservation, they noticed the lone horse looking intently at something out in the field, and a herd of cattle in another corner displayed a similar interest, making moaning sounds. As they drove on, they spotted the focus of attention, and seeing something large walking away from them, at first thought it to be another horse. Closer scrutiny convinced them otherwise, as the big creature was walking on only two legs. Gordon got out his high-powered rifle and scope, and watched it for three or four minutes. It stopped its fast gait to pick up something from the ground several times, but it never looked back in the direction of the car.

Gordon told me that he could easily have shot it, but he couldn't bring himself to kill something that had such human-like features. It was apparently very powerfully built, standing between seven and eight feet tall and weighing between 500 and 600 pounds. It was a dark color, but as it walked swinging its long hairy arms, he noticed grey on the inside of its elbows.

There was no gate to give them access to the pasture or they might have been able to get closer to it. They agreed not to tell anyone about their experience, but when they thought of the children living in the area, they decided to go to the family in whose direction the creature was heading. The man there told them that something had made the dogs very upset earlier that morning, as they had been desperate to get into the house. When he had gone outside to check, he had heard something in the bush and noticed a powerful smell.

About a week before this sighting, Gordon Jr. and his wife had driven along a road on Peonan Point, a long peninsula reaching south into the north end of Lake Manitoba. They saw something unusual ahead of them, and when they got close enough to see that it was something on two legs, it had taken off into a narrow bush. Gordon ran after it, hoping to get a better look as it entered the open field, but by the time he was out of the woods, the creature was already far away.

Gordon Sr. told of an experience his grandfather had many years ago. He and his partner had taken their dog teams and gone north one fall for trapping and hunting.

Before they returned for Christmas, they hunted for moose along the way. However, they came across some huge footprints in the snow, and decided to follow them. They noticed that the creature had stopped abruptly and gone off in a different direction. What they saw when they followed convinced them that it had smelled something to eat and had side-tracked to where a bear was hibernating. Blood indicated that the bear had been killed and then carried away.

A similar story was told to me by a former student from Little Grand Rapids. When he was serving in the Lettelier area of southern Manitoba as a police officer, he and his partner had encountered a pair of Sasquatch one night feeding on a deer which they passed back and forth as they were standing. There is no doubt that these creatures are quick and capable of catching deer. One man, who had been observing a Sasquatch for many minutes as it was sitting in a ditch one evening, said it suddenly bounded across the road in front of him "faster than a deer." He also said that its muscular arms looked like it could "pull a tree out of the ground."

Perhaps the deer that were fleeing from the Sasquatch in the story I got from the young man on the Birdtail Sioux First Nation were doing so because of a fear based on experience. Perhaps they had witnessed some of their own being caught by the creature, and now were desperate to get away. Had this never happened, however, they might just have remained curious onlookers like the horse and cows that watched a Sasquatch cross the field at Fairford.

A man from Fairford, traveling from Gillam to Thompson, also saw a Sasquatch sitting near the highway, its hands occupied with some activity. During the few minutes that he watched the orangey-beige creature, it turned its neckless body twice to look at him. That was enough to "put the fear into me," and he left the scene, not caring to see what the beast might do next.

John, a trucker from Nelson House, had a similar sighting, but his involved three such animals along the highway towards Leaf Rapids. Only one was sitting, seemingly immobile, with its back turned, but the other two stood very close together, face to face. As he watched them, only the one facing him would occasionally move its head to one side to look at him past the other, and then resume its face to face position with no movements whatsoever until the next peek. Finally, after watching them for many minutes, and needing to get on with his job, John honked his horn to see what reaction he would get—and there was none, so he finally drove off.

John saw two other Sasquatch on another occasion, unless, of course, they happened to be the same ones. It was along the same highway in the spring of the year when there was still ice on the lakes. He could see, far out on the lake, one Sasquatch standing beside open water, gazing down at another that was in the water. No real action seemed to be taking place, so he concluded that one must have fallen through the ice as the pair was crossing the lake. He would have watched longer to see what happened, but the voice on his radio ordered him to get moving.

His brother almost hit one on Highway 6 near Devil's Lake recently. He was returning to Nelson House from Winnipeg late one night when his headlights picked up a huge hairy form that was just beginning to cross the road. He had to bend down to see its full height through the edge of his windshield, and he said that if he had slowed down at all, and if the creature had kept up its pace, the two of them would most certainly have collided.

I am surprised that the killing or maiming of Sasquatch has not become commonplace on our highways. But then, very few bears or cougars meet their end because of motor vehicles either. It seems that the more intelligent creatures manage to avoid being hit, but there are some close calls. This was the case not many years ago when a minister from Jackhead and his wife were returning home after services near Waterhen, about 40 miles west of Gypsumville. As they were traveling east around sunset, suddenly something unseen to both of them came out of the ditch and bumped into the driver's side of the car. The wife, who was driving, stopped the car to see what had hit them, and when she looked back through the settling dust, she saw a huge hairy form standing on the road. She turned the car around to go back for a closer look, but the creature was gone. The car was still drivable, but it was written off thereafter. Some hairs were apparently recovered from the side of the car, but I do not know what became of them.

A trucker hauling lumber down Highway 6 to the States admitted, at a local coffee shop, to almost hitting a Sasquatch near Devil's Lake. This highway boasts a good number of sightings, and the ones that I hear about are probably just a fraction of the real count. The road north of Gypsumville, or St. Martin Junction, has not yet been in existence 50 years, but we will never know just how many of the creatures have been seen there during that time. Let me list a few more that I have heard about.

One summer night in the mid-1990s, Ed Krahn of Ashern, Manitoba, was traveling north to Grand Rapids where he worked at the Hydro Generating Station. He had just negotiated the curve on Highway Six at Long Point when he noticed a large, hairy, bipedal figure on the shoulder of the opposite lane, coming in his direction. Immediately the creature lifted up its left hairy arm to shield its eyes from the bright headlights, turned abruptly, and walked into the ditch just a few car lengths away.

Ed admitted that he had not the slightest inclination to stop or slow down, as his body experienced a reaction that he had not had before or since. His hair stood up, and his whole body felt weird. "I couldn't even have turned on a camera if I had had one." He kept looking in his rear-view mirror for several miles.

The next day at work he shared his experience with several co-workers who did not laugh or question his story, but accepted it as just another account of a Sasquatch encounter, like so many others they had heard, living as they did in prime Sasquatch country. Mr. Krahn noted that the creature was seven to eight feet tall with long, coarse, medium-brown hair. Having seen numerous bears before, both on all fours and upright, he was adamant that what he saw was definitely not a bear.

A man now living in The Pas told me he was hitchhiking from Winnipeg to Thompson in the early seventies, about ten years after the highway was built. He got dropped off at St. Martin Junction, so he began walking north, hoping to get another ride soon. After a number of hours, and no vehicles, he spotted a garbage receptacle in a distance, and something big on the road well beyond that. It kept coming his direction as he continued towards it, and by the time he reached the container, it was only about 150 yards away. Although it appeared to be a two-legged creature, he still assumed it was some known animal, but not having had a side view of it, he wasn't sure what to think. When it turned and walked into the ditch, he realized that it was definitely not what he expected it to be. It was as tall as the highway sign beside it, and of a brown color. That sight clinched his decision to bolt back in the direction of St. Martin, where he spent the night in the washroom of a gas station rather than taking his chances in the wilderness.

His co-worker shared a much more recent sighting which, unfortunately, ended in mystery. The man and his wife, in the summer of 2005, had been riding their motorcycle in the Wekusko Lake area on their way back to The Pas from Thompson. Coming over a rise in the road, they saw a very broad-shouldered form "that would make the rear end of a moose look skinny" moving across the road ahead. As they drove down into a dip, they fully expected to see it as they came back up, but they were shocked and astonished to see absolutely nothing less than ten seconds later. Since the wife was not inclined to stick around after what she had seen, they drove the three hours home, where her husband exchanged his bike for his pickup, and returned to the scene. He found no evidence to corroborate what he had seen, but instead of giving up at that point, he returned to the spot the next day with three friends who were also hunters. Again, they found no tracks in the ditch or along the edge of the bush.

However, on hearing, some time later, of a broad-jumper who cleared 26 feet, he realized that the huge creature, aware of the fast-approaching noisy motorcycle, could easily have cleared the ditch and disappeared from sight. But he did wonder about allegations that attached mystical powers to the creatures.

This gentleman also related an incident, among many others, where an acquaintance of Native background had canoed around a point of land and spotted a big hairy Sasquatch waist deep in the water. When he reached for his rifle, his aunt grabbed the barrel and said, "We don't shoot those things."

A lady whose house faces the casino in The Pas described to me what she and her daughter saw late one spring night shortly before the area was cleared for the building of the casino. Although she had gone to bed, her daughter was still on the computer when she noticed a movement outside in the bright moonlight. She awakened her mother, and together they gazed at the strange sight. A big, tall, long-armed creature was standing near some trees, its upper body swaying back and forth "like a pendulum." Then they noticed another one, slightly smaller, standing motionless, which they took to be the mother. Near her was a small one that was moving around as if it were curious. The lady admitted that, although she badly wanted to get a better look at the creatures, she didn't dare go outside. Next morning, however, several of them went to see what evidence they could find, and sure enough, there were three different sizes of footprints clearly visible in the soft ground.

However, this was not her first experience involving a Sasquatch. In the late 1980s, she accompanied her mother on a trip to Thompson. In the vicinity of Paint Lake, something very tall stepped onto the road and looked in their direction. Then it cleared the highway with two strides and was gone. It crossed where a sign stood, towering above it, making it look "like a toy."

Landry Lake is beside the road that leads from The Pas to Moose Lake, and it has gained the reputation of containing several unusual creatures. However, the one that was observed there in the mid 1990s was not exactly an inhabitant of its waters, and if I were to give the following story a title, it would be, using the words of the witness, "The Sasquatch that understood English." I finally met Mrs. Smith in the fall of 2009, and got the details as she remembered them.

Her family camped beside the lake in the parking lot of the boat launch. Early in the morning Mrs. Smith became aware that her young son was not in his bed, so she went outside looking for him. She found him fishing along the shore, but talked him into coming back to bed where he would be safe from the "bear" whose awful odor they smelled as they walked. A few hours later, she awakened and went outside for a smoke, but heard a splashing of water at the edge of the lake behind the trees. Curious, she walked gingerly in her bare feet across the clearing until she could see what was making the noise—and there, a short distance away, was a big brown Sasquatch standing more than knee deep in the lake, pouring water over itself with its hands, alternating one after the other. It noticed her almost immediately, and cleared the distance to the bush in four or five big steps.

It was almost as if it had heard mother and son discussing the terrible smell, and decided to take a bath—or rather, a shower.

The experience did not prevent them from returning to the area, although a garter snake scared Mrs. Smith away from it on a subsequent visit. Her husband joked that a huge Sasquatch didn't bother her, and yet a little snake scared her away.

A couple was traveling from Thompson to Winnipeg around the turn of the century when they saw a large form move onto the highway. Thinking it was a bear on its hind legs, they watched it run into the ditch, its arms swinging, where it hunched down in some tall grass. When they stopped opposite it, the man got out of the vehicle to get a better look, whereupon the creature jumped up and swiftly moved into the trees, momentarily peering back through an opening with curious eyes. Felix gave me the typical description of a seven-foot, muscular, round-shouldered, hair-covered beast.

One of the first stories Paul Shabaga relayed to me years ago was of an incident involving a Sasquatch near Split Lake in northern Manitoba. I was recently able to confirm the details of the account from family and friends of the unfortunate victim.

Apparently a group of young people drove to a river a short distance from the community one summer evening and spent some time drinking as darkness fell. One

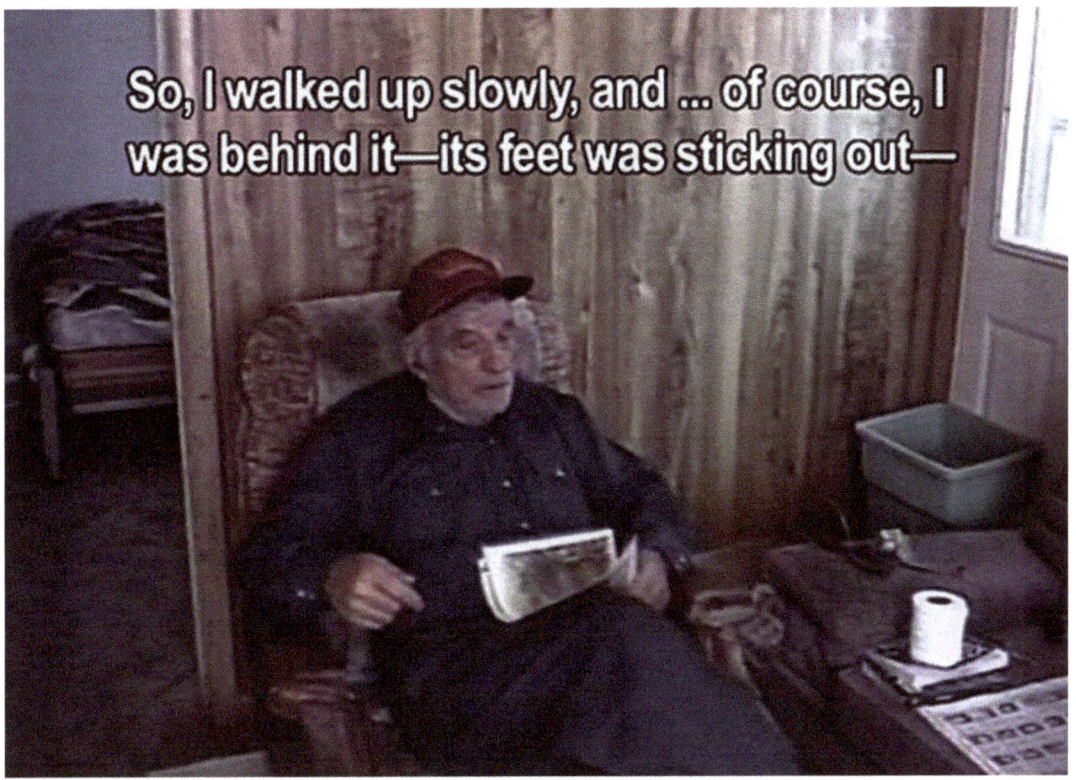

Video still of an interview with Paul Shabaga

man went off to the edge of the bush to relieve himself, and soon returned crawling on the ground. A relative helped him into his truck and drove him to the local nursing station. The man was severely shaken by the experience, and it eventually became general knowledge even though it was not discussed openly. It seems that a big hand touched him on the shoulder, causing him to collapse in fright. Fortunately he was able to crawl back to the group, even though there did not seem to be any real danger. Curiosity was the obvious motive on the part of the creature, which in this case did not exhibit the typical scare tactics towards intruders. There had been no screams, foul smells, or projected missiles, which commonly accompany such an encounter.

Many people tell of seeing fearsome faces looking into their bedroom windows after dark. One man said that he always closes the curtains of his home now, "because I don't want to see that face looking in again."

No one seems to know for sure where they come from or where they are going, but it is often in spring, early summer, or fall that sightings are reported in our Interlake communities. Perhaps they enjoy spending time where they can observe human activity, from a distance, or as Peeping Toms. When they are on our turf, they generally seem to respect our space, but when we are on theirs, that is often an entirely different story—depending, of course, on their particular personalities and moods. It seems that men generally encounter more aggressive behavior than do women or children, almost as if the creatures instinctively know that men with their guns are more of a threat. The girl who was helped home by one of these creatures after being lost in the wilderness certainly has a high regard for them, and this is how I came by her amazing story.

I recall Paul Shabaga, on one of his frequent visits, sharing a unique story with me that he had picked up at the café next door. He had been loosely introduced to a lady by her daughter as having seen a Sasquatch. The tale essentially alleged that a Sasquatch had led her, when she was a young girl, out of the woods to safety after she had been lost for a number of days. Paul's brief encounter with Carrie had not even left him with her name, but he understood that she had grown up in the small Native and Metis settlement called Dauphin River on Lake Winnipeg not far away.

I kept the story at the back of my mind for many years, not expecting to have it confront me again. But it was a reference to Carrie's more recent Sasquatch sighting that inadvertently put me in touch with her, affording me the privilege of gaining her story that she had, for the most part, kept hidden in the back of her mind. I did not know at first that I was unlocking a deeply emotional memory, but when she admitted that she had cried after our first brief meeting, I realized that I was stirring up feelings and memories that had lain dormant for many years. Uppermost was her sense that, had it not been for the creature in the forest, she would not likely be around today to tell us her story.

It happened in the early sixties when Carrie was 11 or 12 years old, just after the highway was built that linked her remote community of Dauphin River to Gypsumville and Highway 6, before the airstrip was completed. It was late in the summer when the berries were ripe that she was carrying out her usual task of bringing in fresh water from a little creek in the swamp at the edge of the forest. Having filled her pails, she set them down, and decided to pick some cranberries, an activity that drew her deeper into unfamiliar territory. When she realized that she was lost, panic gripped her, and she began running desperately, crying with fear, stopping only to rest.

Even as darkness fell, still she pressed on, hoping to gain the edge of the forest. Branches scratched and poked her, and soon she learned to put one hand over her face and feel her way with the other, her eyes being useless in the darkness anyway.

Her biggest fear was of the animals that lived in the bush, especially the bears which were very common around the community. When she became exhausted, she would sit or lie on the mossy hummocks, sleeping fitfully until she sank too far into the softness and got wet.

Carrie recalled being considerably weaker on the second day, but she still ran when she could, crying in desperation and fear, feeling that she was going crazy. Even though it was daylight, the tall spruces that surrounded her made the forest seem so dark since she could get only glimpses of the sky. She did not consciously look for food, but reached out for berries that happened to be nearby. She saw rabbits and grouse, but they meant nothing to her. She only feared bears, and they were constantly on her mind.

She remembered her father telling his children about listening carefully if they were ever lost, but all she could hear was the sounds of the grouse.

More days and nights passed, and Carrie grew steadily weaker. And then, one night, what she dreaded most materialized when she heard the sound of something big lumbering through the bush towards her. She thought instinctively of bear, as it was the only large species of animal she was aware of that might have any interest in her. Its approach felt like a sentence of death as she braced for its imminent attack.

Trembling violently, she addressed the presence that she could not see, remembering the stories of her people. "Muskwa," she said tearfully in her native tongue, "I know you are going to eat me, so do it fast, and start with my head." She addressed it as "Mushoom," or grandfather, telling it that she was lost, that she was going crazy, that she was too small to fight back, and that her parents would never see her again.

The dreaded attack never came. The creature shuffled around her, stopping now and then, its overpowering stench almost making Carrie sick. "It smelled like little puppy poop, or roots freshly dug out of the swamp."

It neither came too close, nor went away. It just stayed nearby. When she realized that it was not immediately attacking her, she reached out into the darkness and felt its hair—long hair that reminded her of the beaver. Still it did nothing but walk around. She clutched a tree for support, and when she finally let it go, the creature bumped her so that she almost fell. She wondered if it was waiting for the right moment to eat her, so her fear and trembling continued. A second time it almost knocked her over, so she moved away from it. If it was on her left, she would go to the right. If it was on the right, she would go left.

Hours went by thus, and when daylight approached enough for her to see its form, she tried to hide from it, only to see it peeking at her. The last time she saw it, it was a short distance away, tall, hairy, and with long arms hanging by its side.

That is when she noticed the stumps. Stumps of the trees the men of Dauphin River had cut in preparation for the building of an air strip. She knew then where she was, so she walked on until she reached the highway that led into her community.

A truck came along, and she recognized a man and his son. She cried as she hugged the old man, whom she addressed as grandfather also, and told him that she had not been eaten up by a bear with long arms that looked like a monkey. The older man had laughed and replied, "Grandchild, you were blessed to be saved by the creature." He went on to say that bears don't do the kinds of things that she described.

Back at home, her bruised and bloodied body told of her awful ordeal, and the pails of water that she had left behind now made perfect sense.

Her parents were skeptical about her story, however, and her sisters teased her about it, calling her "the girl who touched a bear." In time, though, her father told her his own stories, as well as those he heard from others. The one that he would now have identified with had apparently taken place when he was a boy, and it involved a toddler who also had gotten lost and disappeared. The boy had reappeared in the community about ten years later, unable to communicate, and wearing bark. People had been afraid of him at first, but he was cared for and learned to talk. He claimed that big monkeys had looked after him.

Carrie's dad also told her that some men had once caught and tied up a small, humanlike creature on an island, but it had gotten away. He himself had camped on Reindeer Island in Lake Winnipeg once together with some others, and they had been bothered by something that kept throwing things towards them.

The only other person besides her dad who seemed to believe her story was Oliver Ross, the older man who was the first to see her after she emerged from the bush. He told her that he had been lost once too, and found himself sitting on shore. In the evening light, he had seen a strange creature in the water, and it appeared to be catching fish. When he had gone to check the spot in the morning, he had seen its large footprints in the sand, and found two fish that seemed to have been left for him. He wasn't sure if the creature was a real animal, or whether it perhaps had been sent to help provide for him. Another question apparently pondered by his people had to do with the origins of these strange, upright, hairy beings. They wondered if possibly the animals were the result of rare instances of crossbreeding between, for example, bears and humans. Another community's reference to the Sasquatch as "son of a bear" in translation, seems to convey a similar idea. As unrealistic as this idea may seem, it is not unreasonable for the baffled human imagination to consider such a possibility in an effort to explain the existence of something that shares both human and animal characteristics.

The Einarssons, also of Dauphin River, tell of an experience their uncle and cousin related after a hunting trip to the north of the community. After the men had set up camp for the night, something whose eyes had glowed in the dark bothered them continually, keeping them awake. They gave up on the idea of hunting and struck out for home, but the beast followed them all the way back to the community as if making sure that they went where they belonged. I heard practically the same account from folks at Fairford who told of a particular hunting trip north by horse and wagon to the Devil's Lake area. Again, when the men stopped for the night, some creature had harassed them to the point where they decided to leave and camp elsewhere. The trouble continued, however, whenever they stopped, so they just kept traveling the long way back home.

I gather that the origin of the name "Devil's Lake" north of Gypsumville has its roots in experiences such as this one. The "Devil's Lake" in North Dakota across the U.S. border to the south of us is said to contain water monsters which probably account for its name. I have found that most lakes, rivers, islands or hills that bear names that include words like Devil, Spirit, Manitou, God, or Weetigo, have good reason to be so named.

Back at Split Lake in northern Manitoba, another sighting was forced upon a tow-truck driver from Thompson who was returning from Gillam in the early morning hours in May of 1996. "Jigger" had stopped beside the road to catch a few winks and was awakened at dawn by an unusual, unearthly scream. Looking out the windshield, he saw, about 100 yards away, in the middle of the highway, a huge ape-like creature staring at him. The "apeman/caveman" face shocked him so much that the hair on his neck stood up, his rubbery legs shook, his hands shook, his arms tingled, and his eyes began to water profusely, and continued to do so for the next half hour.

The beast seemed to notice his movements in the cab, and shuffled around as if trying to get a better look at him.

Wanting to get out of there, he started up the motor, and with that the creature took three or four swift steps and disappeared into the woods.

We were sitting at a table that was four feet wide when he told me his story, and that, he claimed, was the width of the Sasquatch's body. He estimated its height to be about eight feet, and its weight 800 to 1,000 pounds. It had plenty of hair on its face, and long hair on hands which reached to the knees. The hair on the upper body was dark brown, but below the knees it had a sandy hue. He described its scream as a cross between the bellow of a bull moose and a bull elk.

Jigger said he refused to take tow jobs in that vicinity unless he took along a passenger, and even so, whenever they would approach the area of the sighting, his eyes would begin to water. Watching a movie about a Bigfoot some years later brought on the same symptoms, including the tingling arms and shaking hands.

A young lady, Trish, who lived in a farmyard near me in the area of the 1979 sightings close to Gypsumville, had a surprise awaiting her when she walked along a wooded fence-line gathering sticks for her crafts a few years ago. She carried a tin pail containing a saw and hatchet, and stopped to cut suitable specimens as she moved steadily away from the yard. Each time she finished cutting or chopping, she would drop the tool into the pail and move on.

She had seen something lying in the next field, just beyond a fence that intersected the one beside her, but with the odd dead cow on the farm, she never thought much of it. Not, that is, until she saw, out of the corner of her eye, some movement which turned out to be highly unusual. A creature was pushing itself up into sitting position with its long arms—and then onto all fours—and then finally onto two legs—and then it walked towards her fence-line in a stooped position as if it were an elderly individual. Its back being towards her, it did not see her until it glanced in her direction just before it reached the trees. Once it was out of sight, Trish said she could hear it walking in the narrow strip of bush, its heavy breathing sounding much like low animal snorts and grunts. She did not feel intimidated at this point, as it had not exhibited any aggressive behavior towards her, so she waited to see what it would do.

It was not long before its curiosity made it approach the edge of the trees from where it could look down the fence-line and see her standing in the pasture, and she in turn, wanting to get a better view of it, stepped even further into the open. It seemed to engage in a peek-a-boo activity of sorts for a little while, but when she heard it bang some stones together a few times, she decided it was time to move along. As she picked up her pail and saplings, she smelled an overpowering, nose-stinging odor, and began backing away. She was not downwind of the creature, but

STRANGE CREATURES SELDOM SEEN

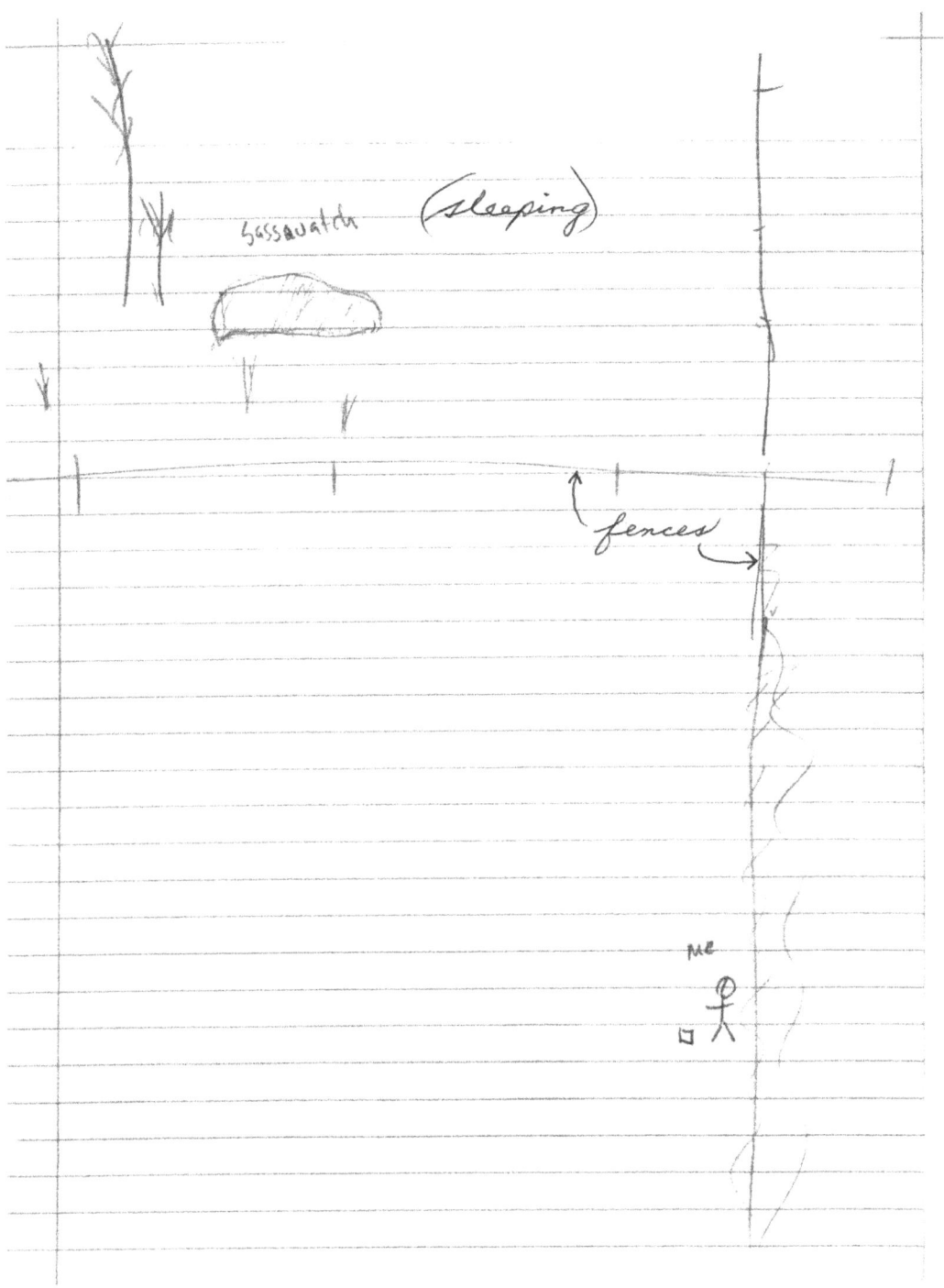

Sasquatch sketched by witness Trish Fontaine.

Sasquatch sketched by witness Trish Fontaine.

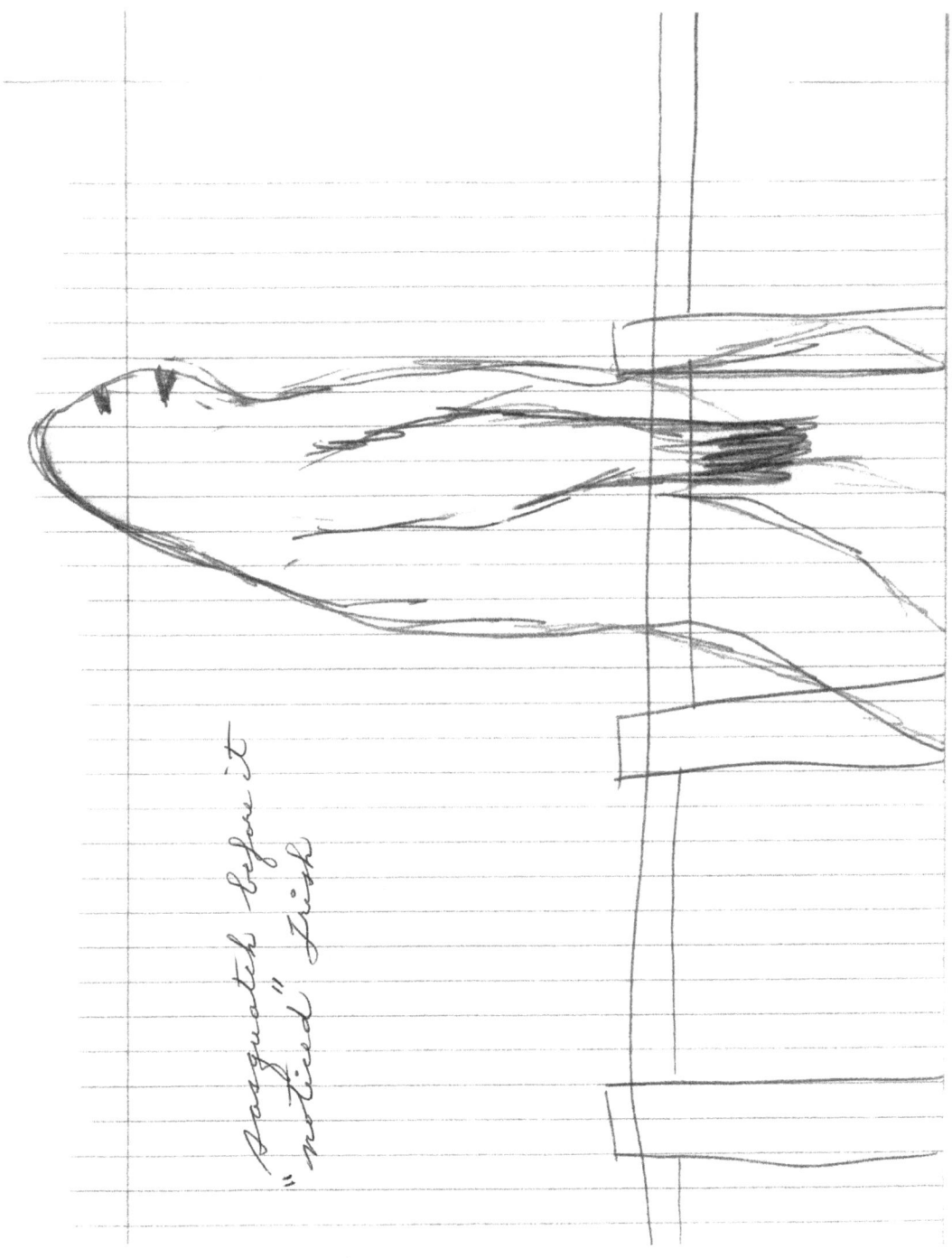

Sasquatch sketched by witness Trish Fontaine.

Sasquatch sketched by witness Trish Fontaine.

Sasquatch sketched by witness Trish Fontaine.

the strength of the smell was so powerful that even the crosswind brought it in her direction.

Again the creature banged rocks together loudly, about eight times in quick succession, achieving its desired effect. Trish turned and began walking quickly away, rattling her pail loudly, and singing at the top of her voice the first song that came into her mind. Every dozen steps she would whirl around to see if the beast was following her, trying not to stare in its direction. With heart pounding and teeth chattering and sweat pouring off her, she survived the long walk home unscathed, leaving the creature behind where she had encountered it.

When she reached the house, the first thing she did was call around for help. Soon two parties arrived, and, taking her camcorder with her, she guided them to the scene of the excitement. They spotted the creature at a distance, its huge hunched bulk looking much like a hay bale as it shuffled around near a road. When a truck sped by, the beast raised up to its full height and ran off, causing a little girl in their group to cry in fright. Since the camcorder batteries were dead anyway, they all packed up and went home.

Thinking back on the days before the incident, Trish told me that there had been some foreshadowing incidents. The night before, their Chihuahua had barked up a storm in the middle of the night, and a few nights before that, they had heard grunting sounds similar to the ones she heard during the confrontation. Strange screams and bumps had also been heard, and a big branch had been thrown against the back of the house. Since unusual sounds persisted night after night even after the sighting, they chose to sleep elsewhere.

In 2005 a sighting was made near a community at the north end of Lake Winnipeg, an event that became known far and wide because it had been captured on video. My son, who lived in Norway House, called to tell me the news and the name of a contact person. So, I talked to the witness's sister, who confirmed that her brother had seen a big hairy creature walking along the shore of the Nelson River, opposite from where he was attending to the ferry. His footage, which I saw later on television, portrayed the typical characteristics of a Sasquatch even though the image was not very clear. One factor that convinced me of the authenticity of the sighting was the sister's statement that her brother was not inclined to go public with the evidence.

Only a few unusual sightings like these get picked up by the media, more so when there are multiple witnesses. The 1979 sightings near Gypsumville, which were published in the newspapers, had more than a dozen eyewitnesses, and the incident created quite a stir. A 2008 sighting in northwestern Ontario near Grassy Narrows by a mother and daughter was also publicized, and it involved a large Sasquatch that had crossed the road in front of them.

Leaving unfamiliar or unusual things alone is a common teaching ingrained in the belief system widespread among the aboriginal peoples. Fear of retribution, often in the form of someone losing his life, is the driving factor behind this cultural trait, so when someone appears to respect those values, then he has, to my way of thinking, demonstrated his honesty and forthrightness.

When I speak with witnesses face to face, or on the telephone, I seldom have any concern as to whether or not the story is fabricated. For one thing, it is usually someone else who has referred me to that person; and for another, I can sense the emotion, attitude, and sincerity of the speaker. Seldom do I need to work at convincing individuals to share their story with me. The way the details invariably match those of other reports leaves no doubt in my mind that the host of different people from a variety of regions and backgrounds are offering corroborating and credible information.

This Sasquatch phenomenon does not seem to be dependent on culture, religion, or race. I believe that more aboriginal people see them simply because they live close to the wilderness areas where these creatures roam, but non-native people living in these same areas are reporting them with increasing frequency as well. And once it becomes more acceptable to admit to such a sighting, I believe there will be an overwhelming increase in reports.

In the mid-1990s, a couple from Moose Lake, which is near The Pas, drove to Grand Rapids using a shortcut through the wilderness since a dry spell made the trail passable. They encountered a big, light orange-colored, hairy monster that sat about six feet high. When it got up and ambled away, it appeared to be between 10 and 12 feet tall. Even its face was hair-covered. They told me that shortly thereafter, they met a forest cruiser who worked for a mill in The Pas, and he admitted to seeing the same creature in that vicinity a few days earlier. The couple spent that night in their truck camper far from where the sighting had taken place, but still they couldn't sleep a wink.

I have heard other rumors of unpleasant experiences concerning folks who spent the night in their vehicles in wilderness areas, and one such story involved two ladies who slept in their car along the Jackhead road in the north Interlake. They felt their car being lifted up, and spotted a large black shape in the darkness.

In the 1980s, after Canadian Forces Base Gypsumville had been closed, leaving behind the residences for use by First Nations families, some boys had walked to the nearby airstrip just before dark. They heard a noise in the bush, so they hid behind some old cars. Soon a big, hairy form walked across the landing field, stopped, and turned towards them as if smelling them, shrieked, and then continued on. Between seven and eight feet tall, it was dark with reddish chest hair.

A man from Jackhead told me that around 1990 he was returning home at dusk from Lake St. George by canoe. At first he thought a moose was beside the water, but when he cocked his gun, the creature, which had been down on all fours, jumped up on two feet and ran away, leaving behind an awful smell. The broad-shouldered creature was black like a bear, he said, and about ten feet tall. He realized that it had been drinking water by lifting it to its mouth with one hand much like humans would. Such behavior was also observed by someone from Cross Lake, in northern Manitoba.

Nelson House has had numerous sightings in recent years, and many of these are of the creatures running across the highway. In one case, a father and daughter were spellbound as they watched two Sasquatch hurry across the road in front of them, one slightly smaller than the other. Further up that highway, between Leaf Rapids and Lynn Lake, a Credit Union manager from the Interlake, and his cousin, were traveling along mid-morning when a silvery-haired creature ran into the bush beside the road. By the time they stopped and went after it, there was not a sound or smell to give it away. It vanished in an area where they were certain that they should get another glimpse of it. The description matched the typical seven- to eight-foot-tall, 500- to 600-pound biped.

An old man from Nelson House who lives in the wilderness took a shortcut through the bush to his camp one day, and came face to face with a Sasquatch. Apparently he remained on the spot, stunned, for a day, and then spent the next two days at his camp, staring blankly into space.

I showed a man from Easterville some sketches of Sasquatch that others had made for me, and he proceeded to make one of a face and head that had long hair extending in all directions. He told me he was driving for his parents in 1976, and had just turned off Highway 6 onto the road leading to Easterville when on the road in front of them stood a tall, dark creature with no neck, long hair hanging from everywhere. The eyes appeared sunken, but with the abundance of hair, the face was barely visible. After staring at them briefly, it walked through the ditch towards the bush, swinging its knee-length arms, and paused at the edge of the trees. It turned its whole body in order to look back at them, and then entered the woods, parting the branches with its hands. His dad, who had heard of the creatures, but had never seen one before, remarked, "That's no man!" Apparently he stopped drinking after that experience! A number of others, including some school teachers, sighted Sasquatch in the area around that time.

Two boys who were urinating off the steps late one summer evening near Portage la Prairie were shocked to see a big, hairy creature with long arms, no neck, and an awful smell standing nearby. Years later, one of them told me that it made him uncomfortable all over again just telling me about it.

STRANGE CREATURES SELDOM SEEN

West of Brandon, a mother and daughter told me that they had been out walking after dark when they saw, in the light of a street lamp, a big form pick up a dog that was barking at it, and hurl it towards the light, knocking it out. A bad smell apparently was noticed after the incident.

In June 2005 a Sasquatch sighting at Fairford was brought to my attention. It involved a young mother who had gone outside for a smoke on the back steps one morning after everyone else had left the house. A couple hundred yards away she spotted what she thought was a root digger, and as she watched, it turned its body and looked at her. That is when she became frightened and went back into the house. She peeked out the bedroom window and spotted a second creature doing the same thing as the first one—bending over, picking handfuls of something, tossing away something. Then they sat down, back to back, and continued working with their hands and eating. After about 15 minutes of this activity, they got up and moved on. She told me that the animals were bigger than men, and had "egg-shaped heads," i.e. pointed on top. They were a dark color and swung their arms when they walked. They had short necks, bent knees, and turned their whole body instead of the neck. Noses were not visible, but she saw the eyes of the one that looked towards her, and that gaze was enough to put extreme fear into her. When I went to check the area where she had seen them, I saw that the ground was covered with a lush growth of a variety of plants.

Two Fairford men were looking for deer or moose early one summer morning, driving through the wooded wilderness east of Moosehorn. The passenger, Anthony, told me that he typically watched the little clearings on his side of the road in case an animal was standing there. On this occasion he spotted a huge, hairy, biped standing near the road, and the sight of it made his hair stand up, and cold chills go down his back. He had not even been able to tell his partner about it immediately.

One boy from a nearby reservation, who told me he saw the creature from fairly close up, said the smell from it was so strong that he almost choked.

If you want to see a Sasquatch, go for a ride in the park—after dark. That's what some young people discovered on September 4, 2006, at 12:30 A.M. in Riding Mountain National Park. They were near the buffalo compound in the area of Jackfish Creek when they smelled something awful, and then saw what was responsible for the smell. In their headlights, less than 100 feet away, something hairy and muscular with wide shoulders and a cone-shaped head was standing beside the road, breaking apart a dead tree. As the occupants of the car alternated between stunned silence and hysteria, the creature bent down only slightly, and placing its left hand on the ground, picked up a knotty little stick with its right hand and flicked it under its arm towards the car. It then ambled out of sight, leaving the car with its terrified occupants alone, some of whom didn't dare to take a good look at the rare sight.

The driver, who was one of two sisters present, paid careful attention, and listed some of her observations: Green eyes reflecting in the headlights; eight-to-ten-inch-long hair hanging from long, muscular arms, colored coffee and cream, or salt and pepper, with silver streaks; a flat, gorilla face with a flat nose and hairless brow; muscles rippling beneath the hair; legs looked like pillars, or tree stumps; no hips; almost like the creature in the movie, *Harry and the Hendersons*—only broader; where exposed, the skin was black; fur was on the back of the hands; it could easily have thrown a log at them, but instead threw a little stick. Her sister described the smell using the words *wet dog, sewage,* and *fart*.

Their dad went a considerable distance the next day, from the Ebb and Flow First Nation to the site in the park, to check it out. He retrieved the little stick (which everyone believed hit the car!) and measured the footprints at a good 17 inches, and the stride about four and a half feet. He also saw where the dead tree with grub tunnels in it had been broken open and branches of other trees nearby had been snapped up high.

He told me that once after he shot a deer in the Duck Mountains he heard such an unusual, loud scream that he didn't bother to retrieve the deer. Someone else that he talked to about it told him that he had given up hunting in that area on account of the unwelcoming behavior of something that tended to throw rocks and sticks towards his camper.

A few years later when I was hoping something similar might happen to me there, I was disturbed by nothing but geese. Not even the smell of the sardines which I had hung high up in a tree was sufficient to lure any creature to me, so I had to leave the next morning disappointed . . . but perhaps a little relieved? I knew full well that if the thing I wished for would materialize, I, scared-of-the-dark solitary me, would not find my little Jeep a very comforting shelter. Were it not for the fact that the extent of their violence against mankind consisted largely of pushing against vehicles or lifting them up, I would not have endured the suspense.

If I had heard one other account from the same mountains earlier, it would have reassured me further that these huge creatures are generally harmless and need not be feared. It was actually in the fall of 2009 that I spoke with Harold from the Sagkeeng First Nation. He told me that he and a friend were hunting for elk in the Duck Mountains in 1999, and met with success late one afternoon. Harold had left his gun beside the carcass, jumped on his ATV, and headed over to where his friend waited so that they together could get the meat back to their camp. Coming over a rise on the trail, he suddenly found himself face to face and only 25 yards away from an eight- to nine-foot-tall, bushy-haired, brownish-grey biped which returned his stare briefly before disappearing.

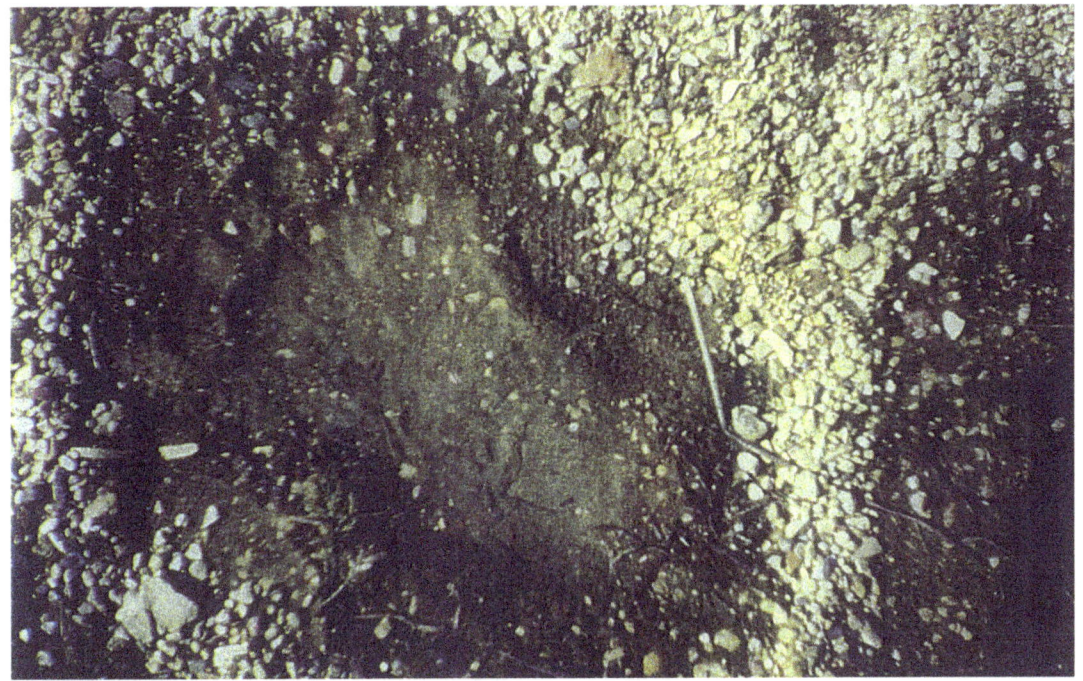

Sasquatch Footprint

Sasquatch Footprint

He did not tell his partner about the encounter, who nevertheless noticed something strange in his behavior, and kept asking him, "What's the matter with you?" Whereas the meat would normally be stored close to the tent to discourage wild animals, Harold insisted that the carcass be placed about 100 yards away. He told me that his thinking was, "If that creature wants the meat, he can have it!" He went to bed with his gun and knife close beside him, and with more than the normal amount of whiskey in his belly.

In the morning when his partner suggested that Harold pick up the meat while he did something else, Harold refused to go alone, prompting the question again, "There is something wrong with you—what is the matter?" The reply had been in the form of a question: "Would you believe me if I said I saw a Sasquatch?" To which the partner had replied, "Yes, I would, because of who you are, and how strangely you've been acting."

Harold informed me that even if he had carried his gun on the ATV that day, he would not have shot the creature since it showed no aggression towards him. Mutual respect existed there—but he never went back to hunt there again!

One hunting guide from the Interlake who heard the creature's voice described the volume of its scream as that of two bulls put together, and when he saw its outline at dusk, it was like two men standing side by side. As it had crossed a field, deer fled before it, making their typical fear-induced blowing sounds.

I have heard of a few sightings that were made from airplanes, and these are either along shorelines, or in muskegs where few trees can hide them. Robert from Split Lake told me that what he at first thought was a big bear in an open muskeg turned out to be a huge two-legged creature that stood up and ran with amazing speed towards a bush.

On another occasion, he and a partner had been working on the ground north of Split Lake near the Churchill River when they heard sticks breaking nearby just behind a knoll, followed by long, loud, eerie yells which seemed to reach "from horizon to horizon," causing their hair to bristle. That was when they saw two trees being pushed over in opposite directions, the ultimate indicator of a furious Sasquatch that did not appreciate intruders.

This episode is similar to one where a hunter saw such a creature shaking a tree as if in anger over his presence, and that of his dogs, who were normally fearless and aggressive. On this occasion they cowered pitifully.

Another muskeg sighting involved a lone Sasquatch standing waist deep in the bog. A man and his young nephew were checking traps one spring, and thought at first it was a moose in the water, but as it straightened up they realized it was a very big Sasquatch that appeared to be feeding on some vegetation that was below the matted surface. As it walked towards solid ground, it seemed to use its long arms to part the heavy growth.

Children and young people riding on ATVs occasionally report seeing Sasquatch as well. One such incident took place here in Fairford a few years ago, and a similar incident took place near The Pas. Further north, at Cranberry Portage, some boys had spent the night in a tree house they had built, and in the morning they saw a Sasquatch sleeping on the ground nearby.

Next to eyewitness accounts of Sasquatch, their footprints are frequently seen as well. In fact, collectors from far and near have photographs and plaster casts that number in the hundreds. Some are much smaller than others, but all of them depict a broad-footed, bipedal creature with considerable weight. The largest cast that I saw with my own eyes was in northwestern Ontario some years ago, and it was either 22 or 23 inches in length. It was a gigantic print.

Most footprints here are seen in relatively soft ground where impressions commonly reveal a foot with five toes, except where deformities appear to be present. Prints in the snow, however, are somewhat rare, and when they are seen in our central regions, where winters are long and severe, it is usually in the fall or in the spring. Everyone I have spoken to on the topic is of the opinion that the Sasquatch either hibernate during the coldest portion of our winters, or hunker down where travel is not a necessity. Caves are generally suspected as the likeliest of dens, but just where they are located remains a mystery. I have yet to hear of someone finding one in a den, but I have heard of several spots that hunters give a wide berth.

A recent incident that took place in January 2009 would appear to contradict the above-mentioned hypothesis, but the sighting of a Sasquatch on a bitterly cold, midwinter day is definitely a rarity. I was not expecting much of a story at all when I was referred to a certain house where an older gentleman, whom I hadn't seen for a good number of years, was said to be visiting his son. The man had been traumatized years earlier by a close encounter with a lake monster that had lifted its huge neck and head out of the water right beside his boat, so I was keen on chatting with him again. However, the casual remark that his grandson had seen a Sasquatch left me with no great expectations at the time, and yet, almost two hours later when I finally left the house, I realized I had become privy to one of the most remarkable Sasquatch stories I had ever heard. In fact, as I glanced across the river to where a couple lived who had just won over $50,000,000, I felt that I now shared some of their euphoria over gaining something of great value.

It was a -35° Celsius morning the day that five young people from the Sagkeeng First Nation headed into the wilderness north and east of Pine Falls to look for moose. Dillon, the 19-year-old driver, was the oldest of the group that consisted of three guys and two girls. His 16-year-old cousin, Ben Ben, who gave me a detailed description of events, had remarked along the way that he wished he could see a Sasquatch some day, even though he didn't actually believe in them—not realizing that his wish would come true very shortly—and fill him with regret.

It was mid-afternoon before the group spotted something on the deserted road in a distance. Thinking that its large shape constituted the moose they were looking for, they drove on. The nearer they got, the more suspicious the object appeared, and, when they realized that it was something upright on two legs, they speculated that it was perhaps a man dressed in furs who could use their help. To make a more accurate determination, Ben Ben looked through the scope of his rifle and announced that it was neither moose nor man, but a big hairy female of some kind.

The creature, when they had approached to within 100 yards of it, ran effortlessly through the deep snow of the ditch to stand behind some small trees. It peered at them curiously for a while as it teetered back and forth from foot to foot, giving its shocked audience an opportunity to examine it from a safe distance. The distance didn't remain safe for very long, however. What possessed the beast to check them out more closely no one knows, but it suddenly began "jogging" towards them, causing much crying and screaming among the fairer members of the group. Shouts of "Shoot it . . . shoot it!" were heard above the din, but although Ben Ben could have hit it with a stone by then, he could not bring himself to shoot it. The driver, meanwhile, was spurred to action, and backed up the vehicle a long way to the nearest intersection; even doing that, the creature was still in sight.

Ten months later, as I listened and savored the amazing story, I was curious to determine the teen's reaction to the whole thing. I must admit that I was surprised to hear him say that he wished he had never had the experience. That is also what he kept telling his mother, day after day, after it first happened. She tried to get him to see it as a privilege, but he did not see it that way at all. For the first week, he had neither been able to eat nor sleep, and, even with the passage of time, he was still haunted by what he saw.

He now understood his grandfather's experience better. After the old man had seen a huge aquatic creature break the surface of the water and face him right up close, he had felt his hair rising as he froze momentarily. He had dropped everything, made for home, and never returned to the lake. The experience had left him so ill that it had taken months for him to regain his health, leading him to make a drastic life change. Had a partner been able to share the experience, it might not have been nearly so traumatic. Ben Ben had four others to verify and discuss the event, so he realized that his trauma would have been much more serious had he been alone in the experience like his grandfather.

He, his cousin, and two others returned to the scene of the sighting the following day. The immense tracks in the snow erased any doubts they may have had as to whether or not their experience had been real. Dillon had stepped into one of the tracks with his boot, and it had covered only part of the track. When I placed both of my size nine runners end to end to see if that length approximated what they saw, Ben Ben shook his head and said, "One more." I moved my first shoe into a third position, and that seemed more satisfactory.

He held out his hands to show the width of the tracks, so, having a tape measure handy, I held it up and saw 14 inches indicated. We agreed that an animal on the scale of what they saw would naturally have huge feet, and the height that both cousins estimated was 12 to 13 feet! Fourteen inches was the *length* of the plaster cast from the 1979 Gypsumville print, and that was considered big. The width of the massive upper body Ben Ben estimated at close to five feet, which is not unreasonable considering its other dimensions. After hearing him say that it was a female, as I probed for a bit more information, I commented, "So it had breasts?" to which he simply replied, "Big ones."

What puzzles me most, however, is what that big mamma was doing out there in the dead of winter. Black bears are not seen at that time of year, and neither are Sasquatch as a rule, so this was definitely an exception. Was she looking for something to eat? Or had her mate kicked her out of their den over some dispute? If so, how big might he have been?

Had I not heard the exact same size estimate from a mature couple whose story I noted earlier, I would be shaking my head too. However, I will not bother to mention an even more fantastic size description until such a time as there is more corroborating evidence to support it.

The following story notes a unique characteristic that has not been attributed to this species through any previous accounts, so I was delighted to hear it described. Since the community where it originated was not in my home province, but situated in a rather out-of-the-way region of northeastern Saskatchewan where I had never had occasion to visit, I made a phone call to someone whom I for several years had neglected to call. As is so often the case, the contact person recommended someone else, someone who I "should really talk to" and in the end that turned out to be just perfect.

In fact, Clifford Carriere had two Sasquatch encounters to share, so that doubled the suspense. I found that the first one not only took the cake, but also the meat. This is how it happened:

Clifford and a 15-year-old helper were trapping muskrats in spring in the mid-1980s, and were bedded down in his brother's cabin. He awoke to the low growls of his dog team, tethered behind the cabin, and knew from experience that whatever was disturbing them was very intimidating, otherwise they would have barked. Bears approaching on all fours, he had discovered, would normally cause the dogs to bark, but if they were seen in an upright position, then the dogs would only growl and whimper.

It was the rattling of the roasting pan outside on top of the old refrigerator that prompted him to get out of bed. He decided to take a peek out the back door, from where he expected to see a bear tampering with the roaster, but when he heard footsteps, he moved towards the window instead. Sure enough, a creature was visible—coming towards the window—but it was not a bear! Instead, it was a tall, fur-covered human-like monster on two legs—veering away as if it sensed his gaze. He watched in amazement as it ambled towards the bush—but instead of entering it, a change of mind caused it to turn and retrace its steps towards the cabin. Clifford ducked down, debating whether to go for the gun or the camera. Doing neither, he chose instead to climb back under the covers to watch what might transpire—and he was not disappointed. Momentarily, the huge figure appeared at the window, surveying the contents of the cabin for several minutes. Clifford, in return, studied every detail of its face and hands, and noticed how human-like all its features were. He wanted his cousin, who was in a bed under the window, to witness this special phenomenon as well, so he kept whispering to him to look out the window.

Perhaps it was a combination of sounds that finally woke Angus—the persistent whispers of his cousin, and the movement of the creature's hands at the window. When he finally raised his head, he found himself staring directly into the face of a monster—for but a moment—and then it was gone! Thinking he had been the only one to witness the unusual event, he tried to wake Clifford—who had a hard time convincing him that he had been watching all along. Then Angus succumbed to his initial impulse—and pulled the blankets back over his head—much to the amusement of his cousin.

Clifford tried to reassure the teen that there was nothing to fear, so eventually they agreed to go outside and walk around the cabin to make sure the creature was no longer around—guns in hand, just in case. That is when they got their second surprise. The creature was gone—like they expected—but the roasting pan containing the four ready-to-cook muskrats on top of the fridge had been tampered with all right. The lid was no longer on the pan, but just leaning against it. When they looked inside, there were the four muskrats—with all the flesh eaten off the bones—put back into the roaster!

It amazed the trappers that not only had the creature not chewed up the bones as would be expected of carnivores, but it also put them back where they belonged.

Not quite enough conscience there to leave someone else's food alone, but enough to put it back where it came from!

Particulars of the creature that Clifford found remarkable included human-like aspects in regards to its fingers, teeth, eyes, ears, and almost hairless face. The short hair that covered its body had a sleek, well-groomed appearance. The description was most fascinating and exciting—but it left me with a very uncomfortable, almost nauseous feeling. It just did not match other accounts.

Whereas this encounter had taken place in spring, Clifford's other sighting happened about a year later in the fall as he was clearing his trails in preparation for running his dog team over them. He heard something following him, but it kept its distance—just out of sight. Wanting to determine more certainly what it was, even though he had his suspicions, he decided to employ a bit of trickery. He knew of a little clearing up ahead, so after he crossed it, he hid himself and waited to get a glimpse of what was following him. The creature was too keen, however, to expose itself in the clearing, but Clifford could just make out its massive form behind the leaves, with its two legs more clearly visible where the foliage was less dense. As if sensing the duplicity, the large biped exhibited its displeasure by shaking nearby trees, bending them back and forth, venting some of its emotion while at the same time sending the signal to the man it knew was trying to hide, that his presence was not appreciated in this part of the wilderness.

I asked him if he had been afraid, but I found that he was one of those rare individuals who are not easily intimidated. Once he was convinced that the species of animal following him was one that he had seen before, he was at ease. It probably helped that he had been taught that these creatures were protectors of mankind, and I am sure folks like Carrie who experienced their benevolence would agree.

As had been the case off and on in recent years, Sasquatch again visited the communities near where I live in the fall of 2009, and seemed to hang around for a while. A few residents of both Fairford and Little Saskatchewan had the privilege of seeing it, even though not all of them considered it as such. One individual from Fairford who has mixed feelings about his experience perhaps has good reason to be ambivalent, since his sighting was not only from a short distance, but also when he was on foot. Michael told me that one evening just before Halloween he was walking along a wooded trail near the school at dusk just after street lights had begun to show their initial dim glow. He noticed a bad smell, like a wet dog, so he had looked around. There, in full view, about 25 feet away, stood this huge, hairy, brown monster, its red eyes glowing noticeably in the twilight as it stared back at him. His body went numb momentarily, but he managed to resume walking, and then running, "faster than I had ever run," his heart pounding. He hoisted himself over a high chain-link fence with no problem, and tearfully asked the night watchman for a ride home—where he holed up in his room for the next three days.

Like others before him, Michael said it looked just like the Sasquatch in *Harry and the Hendersons*. He had gone back to look at the spot some time later just to prove he could do it, but indicated that he would not ever be using that trail again after dark.

We both had heard that it had been spotted by others in the same area where he saw it, as well as elsewhere in the community—and in Little Saskatchewan. A carload of early morning deer hunters encountered it near Highway 6 just north of Fairford, and got close enough to see its brown color and all the other typical features of the species. Around that same time, a man who was passing through Fairford on his way north had been anxious to distance himself from the area quickly after he told a local gas station attendant about seeing the creature a short distance back. Another Fairford man told me that it was just a few miles south where he too saw it in his headlights as it crossed the highway. When he opened his window once he was stopped, he not only noticed a foul odor, but also heard grunting sounds as it disappeared into the woods.

About this same time, a similar scenario played out over 100 miles to the northwest of Fairford near the Shoal River First Nation. Instead of only one observer, however, here there were two witnesses, father and son, and they got their view from atop a large hunting stand that they shared. It wasn't that there hadn't been some warning of strange things to come, however, for an incident back at camp earlier that morning had stirred their blood already.

They had arrived at their hunting camp the night before, making the three-hour ATV ride with three others on a total of four machines. They had spent the earlier part of the day hammering away fixing up the camp buildings, but had become uneasy about some noises in the nearby woods. Not recognizing the unfamiliar sounds of sticks breaking and banging as moose behavior, Robert's son Ed took a walk along a trail that circled the area. He admitted to me that at one point he got such a strong feeling of being watched that the hair stood up on his neck. He made it back without incident, however, and rode over to the hunting stand with his dad later that evening, a distance of about a mile and a half.

The pair had been calling moose for a while when Ed saw movement in the direction that he was facing. A massive, dark figure stepped from behind some willows just over 100 yards away, and seeing the men in the stand, immediately backstepped for cover. Its outline remained visible, however, and its legs, which were not altogether behind the willows, were clearly seen. The two men stared at the creature in disbelief for a good while, their guns shaking in their hands. Finally, after the monster disappeared from view, Ed was dispatched to bring their ATV close to the stand so they could load up and take off for camp.

When they arrived, it was not difficult to muster the party for immediate departure, for, in addition to the odd noises they had heard earlier, the boy in their party had also seen something tall walking through reeds that evening. They roared home in tight formation on three machines, leaving the least reliable two-wheel drive behind, stopping only to re-install the chain that kept jumping off one machine. In spite of that, they informed me that the trip back had only taken them about two hours compared to the usual three.

Subsequent trips to the camp later that fall consisted of a beefed-up hunting party that included a generator so that lights could at least give some feeling of security. Robert explained that, whereas they and others from the community would always get bull moose in the area each fall, there had been no sounds or signs of any at all in the fall of 2009. In fact, he knew of no one who had shot a moose up until the time he spoke with me at the beginning of February. We discussed the issue, and wondered aloud if the presence of the Sasquatch had any bearing on the lack of moose. Robert then recalled a time, over 30 years previous to this, when he and his family had actually lived and ranched full time in the area where their camp was now situated. His dad had kept a large number of horses, even though he used tractors for haying, but they kept disappearing and could not be found. Robert remembers that one evening when his dad returned from searching for the horses again, he announced that they would be moving back to the community the next morning, and that is exactly what they did. There had been no explanation for the sudden move, to his knowledge, not even to this day, so we were left wondering if his dad had perhaps found some evidence that connected this same mystery animal with the lost horses. Since horses and moose have some similar characteristics, perhaps history is repeating itself on this isolated peninsula of Lake Winnipegosis.

When I was in Alaska in the summer of 2005, I visited the Native Cultural Centre in Anchorage where I met a lady from Sitka. She told me that her grandmother remembered hunters bringing home a small, very hairy Sasquatch, putting it in a cage, and feeding it raw meat. After someone began giving it cooked meat, it had died.

On that same trip I met another lady who related an experience she and two other women had in the 1970s. They had left the car to answer the call of nature, and climbed a mountainside that was covered with brush and tall grass. She had almost stepped on the foot of a big, brown, hairy creature that was sleeping on its back with its arms behind its head. She screamed, attracting the other two women, but she grabbed them and rushed them back to the car. All three of them apparently fell when they crossed the ditch, and immediately forgot the incident, and therefore never told the three men in the front seat what they had seen. It was about four years later that their memories of the event returned. Because of this, they felt that the creature had some "mind-control powers."

The Alaskan people seemed to have numerous stories of beings fitting the description of Sasquatch that were in the habit of abducting women and children, often releasing the males. Ours in central Canada seem to be more benign, and do not have that same reputation. Here, aggressive behavior towards humans is not reported in very many instances, although the creatures' tendency to scare intruders out of their territory is legendary.

As I traveled from Manitoba to the American southwest in December 2009, I was pleased to hear of Sasquatch sightings in various locales along the way, even in areas that I thought not to be suitable for them. It seems, however, that wherever there is sufficient wooded wilderness for these creatures to hide, they do on occasion pass through. Here at home, however, with endless wilderness to give them safe passage, they tend to visit more frequently with almost predictable regularity—a phenomenon that is most interesting and entertaining, to say the least.

Many folks carry the suspicion that Sasquatch have an unknown dimension to them. Some wonder why its tracks sometimes seem to end abruptly. Others wonder where it disappeared to when they thought that they had it within their reach. Still others seem to have reason to wonder if UFOs drop them off and pick them up again. And everyone wonders why we have so little in the way of definite proof, like clear pictures or carcasses. Even the abundant hair samples seem to be of little value since there is not yet a recognized specimen to make a comparison.

From my perspective, having heard literally hundreds of accounts, most of them first hand, I do not rule out much any more, as I have become keenly aware of forces and powers on earth that I have heard a good deal about. However, the apparent earthiness of these creatures seems to present overwhelming evidence of their physical reality.

Not a few people have had the opportunity to bring down a Sasquatch, but respect for a species that somewhat resembles our own seems to have spared a good number of them. But, if they did not also treat us with similar consideration, it would be a different story.

Reports of Sasquatch having been killed by a variety of hunters in different parts of the continent exist, even if they haven't been verified. Their awful smells and chilling screams have been reported by multitudes, many of whom also saw the source, and those who did, when they attempt to explain what their eyes beheld, essentially all list the same particulars. The only variations I am aware of are in regards to their size, hair coloring, and gender. A few witnesses, however, remember the reflection from the eyes being green, while the majority indicate that it was red.

Something that can be seen, smelled, heard, felt and killed surely is a real being. It may even be that there have been more Sasquatch hit by vehicles than some other animals like the cougar. Besides the minor impacts I have heard about, there are a good number of close calls as well. There is no doubt that we are dealing with an intelligent, ape-like creature with keen senses all around. And with its size and strength, there is not likely another beast on earth that it needs to fear. As for us, hopefully we will be able to enjoy co-existing with these interesting creatures indefinitely.

Whereas I came close to ignoring this creature altogether, I now, as a consequence of a goodly number of vicarious dramatic and enlightening encounters, feel an affinity bordering on kinship for this unique beast which walks among us. Its animal instincts appear to be far superior to most of its peers, while its mortal traits are most intriguing.

I feel I must add some information which has just recently impacted what I have believed about the Sasquatch until now. You may recall statements I have made about stories and descriptions that matched—always matched, giving me the satisfactory impression of conformity and unanimity. Even a few paragraphs back I said, "The only variations I am aware of are in regards to their size, hair coloring, and gender."

Now I must modify that statement due to two recent accounts which exposed a glaring and upsetting deviation from all other accounts as I understood them.

When Clifford Carriere gave me his story first over the phone, and then later in person, he mentioned several features of the creature which made me a little uncomfortable because they didn't exactly match the more typical descriptions. One feature was its short, sleek, black, fur—"like a bear"—as opposed to the long, shaggy, unkempt hair of others.

Another feature was a neck—"just like ours"—which allowed the head to turn from side to side as opposed to other accounts where the animal's entire upper body was involved in the turning of the head.

Then there was mention of the human-like head—not dome-shaped, but more flat like ours; ears that protruded like ours, not hidden under long hair; a nose that was not flat like an ape's, but also protruding like ours.

I must say that my heart sank when Clifford showed me his drawings of the creature he saw. I just did not know what to think for a few months until I was referred to another gentleman whose sighting took place in 1960 near Swan River—which is not that far from Cumberland House!

Archie Motkaluk, I discovered, had just recently gone public in December of 2010 with the amazing story of his encounter that dated back 50 years. At the age of

20, in 1960, he had taken horses and sleigh to a distant forest to get a load of firewood during his visit home from Winnipeg over the Christmas holidays, and found himself sharing the large clearing in the bush with what turned out not to be a human, like he assumed for the first hour or two. As it worked its way slowly but steadily towards him, he saw that it was eating whatever berries were near the ground or on shrubs, and, having heard of the Sasquatch, he realized that he was uncomfortably close to one. It just kept coming until it was no more than ten feet from him. That is when it hissed and snarled and carried on for about a minute, causing him to freeze, with an axe useless in his hand.

When it finally stopped its demonstration of displeasure, the two stared at each other for a number of minutes, after which Archie finally recovered enough to be able to take a step backwards. So did the creature. Another step back—and the creature the same . . . until Archie reached the sleigh and the rifle. Rather than spook the horses or unnerve the Sasquatch, he decided against firing into the air. Not wanting to go all the way back home with a half empty sleigh, he resumed cutting wood, keeping a close eye on his new acquaintance who, he noticed, was also keeping an eye on him as it continued foraging nearby for berries.

When he finally left for home with a full load of wood, the creature remained behind—and that was the last he saw of it. Unfortunately, however, it was not the end of his dealings with it, for within three weeks, after being plagued by recurring nightmares where he was continually confronted by the Sasquatch, he thought he was going crazy. For the next 40 years, the pills that his doctor prescribed were his standby, and not even his wife knew what his nightmares were about. Only his mother knew his story from the beginning, and it was she who encouraged him to sketch the animal he had encountered.

If Archie had not become indignant when he heard someone on television insist that Sasquatch didn't exist, who knows how much longer his encounter would have remained a secret. Not only are there the usual naysayers to deal with, however, but there is also the issue of his atypical description that is causing some frustration. When people expect (as did I) a description to match what is generally understood about these creatures, then the mention of a flat human-like head, ears, nose, neck and arms—coupled with short soft fur and an awkward wobbly gait—all these details make the story sound even less credible.

Whereas Clifford's Sasquatch was obviously not a female, Archie's obviously was. Standing only feet away from it, he did not notice any male genitalia—nor female, for that matter, but there was a slight indication of breasts under its fur. We agreed that it must have been comparable to a young human teen female, as it likely had not yet produced any offspring. It stood only about 6' 2" to Archie's 6' 4", but

STRANGE CREATURES SELDOM SEEN

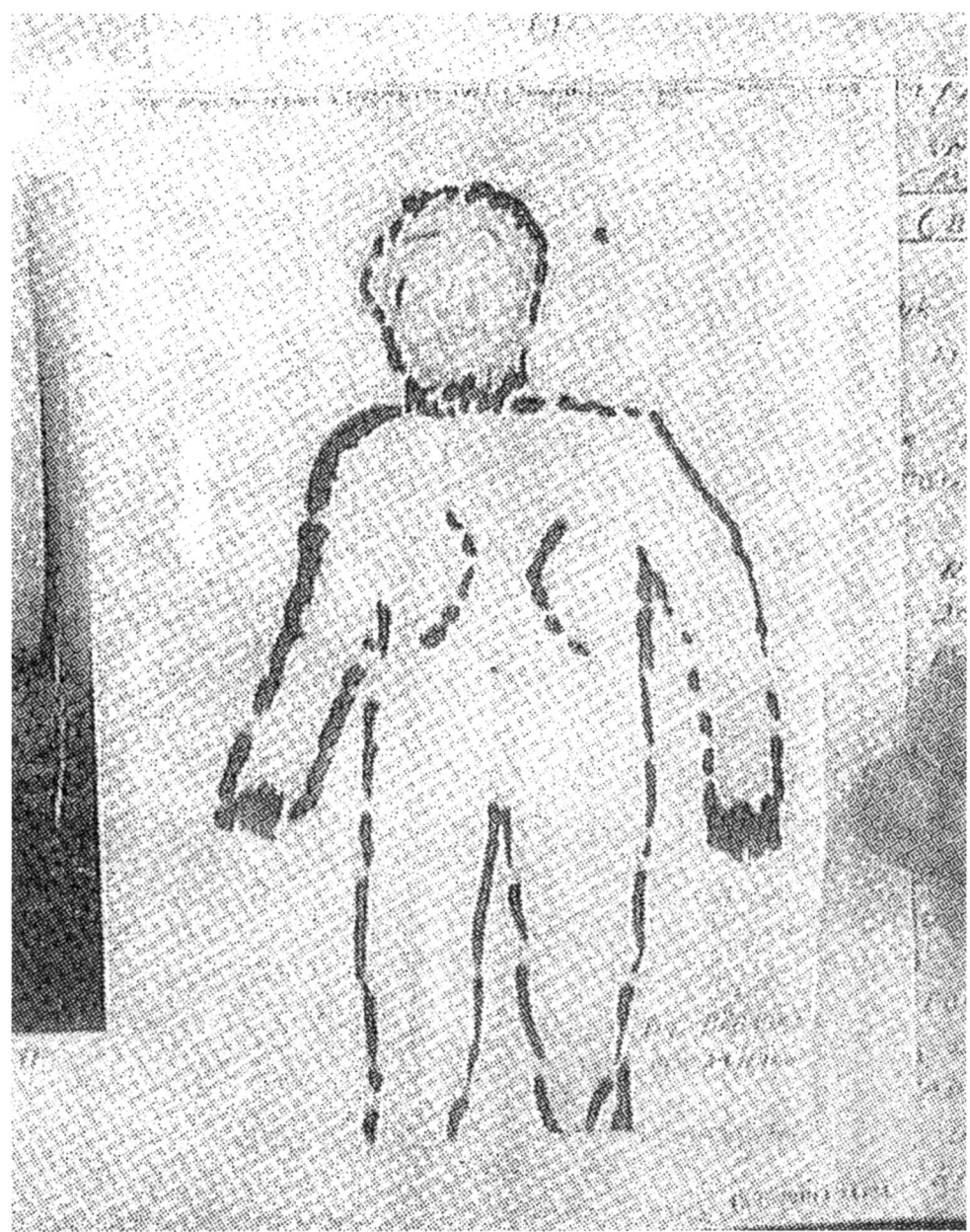

Archie Motkaluk's sketch of the creature he saw in 1960.

The author discussing the "human-like" Sasquatch with Clifford Carriere

he estimated its weight as considerably more than his. He called it a 325-pound hairy woman!

Archie described the fur as being soft and fine "like a cat's," and no more than an inch long over its entire body, giving it a sleek, well-groomed appearance—similar to Clifford's description. When I showed Archie the pictures that Clifford had sketched of his creature's head, I was delighted to see Archie agree with practically all the salient features shown. The frontal head image, however, which shows a round face, he recalled being narrower like most human heads, but with the flat top.

I was surprised that both men agreed on a prominent, protruding, human-like nose—and ears—and a largely hairless face, whereas the more accepted Bigfoot features a wide, flattened, ape-like nose, with some longer, unkempt hair on its face. They also agreed on a neck and shoulders—which really sets the two types apart. The neck allowed the head to turn from side to side like ours, whereas all the typical Bigfoot reports emphasize that it turns its entire slope-shouldered upper body in order to look to the side. The top of their heads are always described with words like "pointed," "dome-shaped," or "egg-shaped."

Both men agreed that the arms were not extraordinarily long, but comparable to human arms. Sasquatch reports always emphasize the extremely long arms that are seen swinging out front and behind as the creature walks.

Clifford noticed that the creature walked swinging its right arm forward as it put its right foot forward—opposite of what we do—giving it a different gait than ours, but fairly smooth, nonetheless. Archie, who was able to clearly watch his animal approach right in front of him, insists that it had a very wobbly, side-to-side movement as it walked, and he wondered, seriously, if it were even capable of running. We agreed that if it were perhaps an injury that caused the awkward gait, it would have had to be one that affected both legs similarly—so we ruled out that possibility.

It is interesting that both men insisted that the creatures' features were quite human. In fact, Archie went so far as to say that he felt his was 90% human and 0% ape.

Both men said that their teeth resembled ours—only they were bigger and not as white! Archie estimated the width of its front teeth as between half and five-eighths of an inch with no fang-like incisors. Because of such large teeth, he did not feel the animal could possibly have as many teeth as we do.

He noted that it seemed to have difficulty picking the small berries off the bushes, but then its fingers, he said, were somewhat "fatter" than a man's. This lack of dexterity perhaps helps to accentuate its animal nature.

Whereas Clifford's animal was black, Archie's was brown. Archie remembers no smell at all, yet Clifford reported an odor which was not, however, overpowering as with most Sasquatch.

Illustration by Alistair Anderson. (From *Grizzlies & White Guys* by Clayton Mack, edited and compiled by Harvey Thommasen, Harbour Publishing, 1996. www.harbourpublishing.com)

Taking all the features into consideration, it appears that the two creatures—the "Clarch" (experienced by Clifford and Archie!) and the Bigfoot are two entirely different species. If the idea were to be entertained, however, that such animals are part human and part something else, then the Clarch certainly would be the better candidate.

Until I shared Clifford's sighting with Archie and explained that both their sightings were shockingly new to me, he was under the impression that he had seen a true Sasquatch. He was frustrated that people found his story so hard to believe—but had I not heard Clifford's account first, I would have had difficulty with it as well.

Many years earlier I came across a fascinating book entitled *Grizzlies & White Guys*, which chronicled the exploits and experiences of Clayton Mack—an expert hunter and guide in the Bella Coola region of British Columbia. Recorded also were Clayton's encounters with some Sasquatch, along with an artist's rendering of the creature.

I ignored that sketch for years, considering it spurious and misleading. However, years later when I managed to bring Archie and Clifford together for a joint discussion and interview, I remembered to bring the book with me. When I showed them the picture of the creature that Clayton Mack had seen on the west coast, both men agreed that it closely resembled the animal that they had encountered as well.

I was overjoyed, to say the least, because finally a major piece of what had initially appeared to be a contradictory puzzle had fallen into place.

So it is with much appreciation to the author of the book, and its publisher, that I proudly include their sketch in my book as well.

Recent rumours of sightings in the same region where the two men had their encounters engenders a new wave of excitement, as perhaps there may be even more corroboration of this atypical Sasquatch, the "Clarch," on the way!

BIG BIRDS AND BIG BATS

THESE ARE NOT UNHEARD OF like some other creatures in the line-up. Reports of huge birds or flying creatures have been around for a long time, but mostly in the more southern and western regions of the North American continent—or in other parts of the world. Stories of birds so big that they tried to take off with children or small farm animals. The ones I hear about, however, are mostly in the north where you wouldn't expect them to be—where the winters are so severe that you wouldn't think any airborne creature, especially those without feathers, would put up with it, let alone flourish in it. There are a good number of birds that do. Jays, ravens, and owls, for example; a host of others migrate to warmer climes as winter approaches.

The big ones, or at least some of them, have been seen in the dead of winter.

Just how big are they, and what do they look like? As of now, all I have to go on are accounts told to me, but a good number of them are first hand.

The most common bird story is one that seems to be ingrained in the psyche of the indigenous peoples all across the north. Legends of the Thunderbird are found everywhere, and they consistently include the natural phenomena of thunder and lightning. The typical thunderbird story that I hear goes something like this:

During an intense thunderstorm, or immediately afterwards, a gigantic bird appears from somewhere in the sky. It descends to the earth where it snatches big snakes from on land or water and flies away with them.

Not so long ago I might have placed this scenario into the category of legend or myth, but not anymore. Although certain aspects of these stories may not necessarily be typical of common experiences, I have found a commonality among witnesses that tends more and more towards credible experiences.

The northwestern region of Manitoba is where the majority of giant bird stories originate. I would be a bit concerned, however, if knowledge of them were confined to this area alone, so I am pleased that entries in my notebook are drawn from divergent places, not only within the province, but also across the country.

The following account was told to me firsthand by a gentleman who a number of people from Pukatawagan recommended. Don Nice had seen a huge "bird" in the 1970s near Puk, before the winter road existed. One time as he was hauling boxes of fish behind his snowmobile, he stopped near the rapids at the Morin Lake portage to straighten out his load when, in the bright moonlight, he saw what looked like an airplane approaching just above the trees—only there was no noise. As it glided swiftly over him, he heard the sound of rushing air even though there was no visible flapping of wings at the time.

A cold shiver coursed up and down his spine as he realized the danger he was in, for the body of the bird was much larger than his. The wingspan he estimated at well over 30 feet, and the distance from nose to tail was only slightly less. It resembled neither an eagle, an owl, nor any bird he had ever seen in his life. It flew low enough that he could have thrown a ball that high. It was a dark grey color, and something orangey under the tail, in the shadows, he presumed was its feet. Its body reminded him of the fuselage of a small plane, about the size of a Piper Cub, with its head "blunt like a frog's, but a little more pointed."

The most curious feature about this "bird," however, was that it was possibly *not* a bird in the normal sense, because it seemed not to have any feathers. Instead, the wings, which were shaped more like a bat's, were covered with a skin-like fabric.

Featherless "bat" seen and sketched by Don Nice.

His father-in-law, who was from Puk, had given Don to understand, through his daughter, that he had seen a huge bird once too.

A similar story, also out of Pukatawagan, takes place after dark as well, leading us to believe that these particular birds, which are never seen in daylight, must be strictly nocturnal. As one man pointed out, "ravens and eagles don't fly around at night." The following sighting, if it can be considered to be one, took place in summer during the firefighting season when some men were in bed on a bright moon-lit night. Around midnight, a large shadow covered the tent, and they knew it had to be a big bird from the swishing sound its wings made. Since it kept circling the tent, one of the occupants, Angus, decided to challenge it, so he grabbed his palaska, or fireman's axe, and stepped outside—but it had vanished.

He remarked that it was not known to bother humans—but on that occasion he wasn't taking any chances.

In the same community, but much more recently, a young band constable stopped by his sister's house one winter evening. She had asked him to feed her dogs outside in her back yard, so he took out some food. A dog that accompanied him cowered around his legs, and the numerous dogs that normally greeted him with loud voices were strangely silent. It was pitch dark and snowing heavily, but he thought he made out a dark form in front of him. Thinking that perhaps his sister had gone out for a smoke, he spoke to her even though he did not see the light from a cigarette. When there was no answer, he reached towards the object, and that movement precipitated a loud flapping of wings as something rose up, branches breaking in its path. Perhaps the creature was just looking for an easy meal, and had to leave disappointed.

A neighbour told me that she had been walking home in the dark one night when a large, dark shadow passed over her. At the same time, a raven let out a squawk—something they don't normally do in the dark.

Some interesting lore held in common in this community is that certain large birds eat the bark of poplar trees. One man described a feeding site in some detail. He had seen a poplar grove where, about ten feet above the ground, a patch of bark about a foot square had been removed from numerous trees. And about three or four feet below that, a set of claw marks had dug into the sides of the trunk. Other trees showed the same evidence, suggesting that something bird-like was feeding on the bark, the scrape marks resembling those made by beaver teeth, but running vertically instead. Chips of bark in the snow were evidence of recent feeding, and the lack of tracks confirmed the presence of a winged creature. The debarked area on some trees had healed over, indicating that this was not just a recent feeding station.

Which bird it is that seems to dine on poplar bark is still a mystery, but the following story leaves no doubt as to what some of them eat.

Two brothers shot a deer one summer near their community located just to the west of Lake Winnipegosis. The driver, who told me the story, said he drove his car to where the deer lay, and rummaged for his hunting knife beside his seat. He heard a strange sound, but was surprised to hear his brother say, "Don't move!" He looked out his open window and saw two huge golden eagles standing beside the deer, one at the head, and the other at the tail. They stared back at him for a long while, and then one of them turned to the other and let out a piercing scream, whereupon the larger eagle jumped onto the deer, sank its talons into it, and flew off with it "like nothing."

The brothers lost their food supply, but they gained an awesome experience. I had no reason to question the veracity of the story, as the teller never approached me with it at all. It was his wife, who, after talking to me for several hours about her giant beaver sighting and other stories, prevailed upon her husband to come and share his experience with me.

This story prepared me for further mind-stretching concepts regarding the habits of large birds, and I came to realize that if a bird that doesn't stand quite as tall as a human can fly up with a deer, then perhaps a much larger bird could as easily make off with something the size of a moose, leaving its tracks suspended in the snow. This is exactly what I heard from a few hunters, and it had everyone puzzled. No one dared to suggest the obvious scenario, that if the tracks stopped, something had to pluck it off the earth from above. Knowledge of the unusually large birds was just not prevalent enough to make the connection—and even if some suspected the obvious, they knew better than to speak of such nonsense.

One trapper from Nelson House did find out, however, and lived to tell about it. Sometime in the mid-1900s he was attacked by a large bird in the area of Harding Lake. He managed to shoot it down, and saw that it had the head and wings of an eagle. His grandson, now an old man, told me that it had screamed almost like a human after it was shot, and added that his grandfather remarked that "it was capable of picking a moose off the ground."

This same grandson mentioned seeing a huge bird that resembled an eagle, each wing measuring about 14 feet, and "a big nest up on a rocky hill."

Also at Nelson House, but only a few years ago, two teenage boys were canoeing along the shore when they scared up a huge bird that was a combination of colors—black, grey, and white. It "made a bald eagle look small," but it seemed to have difficulty getting off the ground. They heard the whoosh of flapping wings even before they saw it, but once it was airborne, it flew across the lake with slow, powerful strokes, leaving the boys in a state of shock.

An elderly widow from the same community shared her experience with me as well. I had been referred to Dolly by a number of folks, but the day I finally met her she was not feeling well. With the assistance of several of her children who were with her, however, she gladly told me of her experience.

It had taken place in the late sixties, just after the road to their community had been built. She and her husband had been picking berries near the highway about 20 miles west of Thompson when they saw, among the shrubs, two big, brown birds. They stood about eight feet tall. Before they flew, they ran a short distance, and when they began flapping their wings, it had sounded like thunder.

I had to wonder, after hearing that, if the name Thunderbird had any relation to the sound of their wings.

One lady from northern Manitoba remembers, as a child, climbing into a big nest together with some other children, and walking around in it. Rocks made up the sides, and there were bones lying inside of it. In southern Manitoba, something that matches this description is advertised on a highway sign near the Lake Manitoba Narrows as a Thunderbird nest. Some local elders think that it was man-made in honor of the giant bird, but others feel that it is a genuine relic from the past. I have heard of two other such nests that were known to be in the same general region. One was in a gravel area east of Camper, and the other was seen at Big Rock, that unusual rock formation at the north end of Lake St. Martin believed to be associated with a meteor strike from the past before glaciation took place.

Alleged thunderbird nest near the Lake Manitoba Narrows.

A number of big nest stories also originate in the area to the east of Lake Winnipeg with general descriptions that include key words like "big nests," "big sticks," and "stones." From handed-down accounts, I gather that some, if not all the nests, were constructed of stones, with the interior "bedding" consisting of sticks—and sizeable ones at that. One youth who told his experience to my informant many years ago said he encountered two such nests near each other, and when he had cleared out the debris from inside one of them, he could barely see over the edge. Another significant observation he made had been in regard to the nature of the rocks used in the construction. They were apparently not the local granite stones of the Canadian Shield, but rather stones "brought from beside the lake," meaning probably the flatter and lighter limestone which would have been "flown in" some distance, worthwhile obviously as a much more stable and safer construction material compared to the heavy, rounded granite.

The material used in the construction of a huge nest that I first heard about in February 2010 when I visited the Skownan First Nation was a practical type one would expect birds to use, rather than one built of stones. In speaking with an alert octogenarian about unusual phenomena, he was reminded of what he stumbled across as he was pushing bush for some American moose hunters many years ago. Inside a spruce bluff was a large nest of sorts, made exclusively of spruce boughs that had obviously been broken off the surrounding trees. Fred said that he had approached the structure with his gun ready, thinking that it was possibly a bear's nest. It was empty, however, with no ground level entrance that would suggest a bear as the architect. It seemed obvious to him that access to this nest was strictly from the air, and therefore belonged to a large flying creature that had recently used it. The sides were about four feet high, and the sweep of Fred's hand indicated that the nest would have taken up most of his living room. He had refused to return to the site when the Americans had been interested in seeing it.

The location of the nest was in the same general direction from his community as is a lake he referred to as Cow's Head Lake. Apparently his ancestors had found a large skull there that resembled a cow's, only larger, hence the name of the lake. Cow-like bellowing is the sound that many have heard coming from around the Waterhen Lake area, and some claim that the sound comes from under the land as well. The term "water buffalo" is commonly used to describe this creature which no living individual seems to have seen.

A member of the same community perhaps saw the very bird associated with that large nest. In the summer of 2009 a dark bird the size of an airplane was seen emerging from a cloud only to reenter it shortly without so much as a flapping of the wings. But then, the vast, uninhabited, boreal wilderness to the north of this

Scenic winter road near Pukatawagan.

Cave hunting summer style.

community, which is situated near the east shore of Lake Winnipegosis, makes for a plausible habitat for creatures wishing to avoid civilization.

Although the odd big bird sighting has come forward from the south-central portion of Manitoba, it appears that they are more frequently seen in the north where there are fewer people to disturb them, and where there are still huge tracts of pristine wilderness. The nests discovered in the more southerly regions seem to have been long abandoned, perhaps at the time when human presence increased.

There is another type of "nest" that I would like to discuss, one altogether different, and to date still tantalizingly mysterious. From my earliest visits to Pukatawagan, I met with Miles and his brother Celestin regarding a cave of sorts that had several strange attributes. For one, the stench emanating from it was terrible. Also, there were many bones nearby, even large ones of moose, including their skulls. The brothers had not dared to go near the opening, knowing that something lurked there that was capable of bringing down a moose, and perhaps even capable of bringing it "home." They seemed to have no idea what creature inhabited this mountainside, and it puzzled me for years, as well... along with the stories of suspended moose tracks. My imagination kicked in, and I wondered what animal was capable of bringing down a moose, and would take up residence in a cave. At first I thought mountain lion, or lion as in the type described to me from the stories of some elders. Then I thought of Sasquatch, another reasonable possibility, as I knew them to be capable of taking down moose, and I had always assumed that they must hibernate in some such shelter.

It was not until I left Pukatawagan from my 2008 winter trip that a more logical possibility came to light. I stopped at Sherridon to meet with the man who not only described to me what he saw in the moonlight in the seventies, but also sent me a sketch of the "bird" that flew over him. Don and I agreed that the wing structure that he had observed and drawn was more bat-like than bird-like, and when I happened to mention that I was trying to search out a cave that smelled badly and was surrounded by bones, he asked the illuminating question, "Don't bats live in caves?"

We agreed that now we quite likely knew what lived in the cave. Everything matched. Living in a confined space, raising their families for generations—all this would account for the smell, and the bones of prey. And doing their hunting by night would make it difficult for humans to know much about them.

I made another winter road trip to Pukatawagan in 2009, and heard of several more cave sightings. Each report mentioned the powerful odor, and bones nearby. Most caves had been discovered in the course of firefighting, when men walked over and around mountains that they otherwise would avoid. Only one individual had tried to enter a cave, and he didn't get very far. A blowing sound sent him running immediately, and when he rejoined his partner, the fire-crew foreman who

Lena Colomb is a grandmother who still hunts and traps, and made this small example of the traditional rabbit blanket.

Searching for one of the elusive caves near Pukatawagan

was just a short distance away at the top of the rocky hill, his boss had exclaimed, "Let's get out of here . . . that thing might have wings." The older man, Adlerd, had had some experiences that made him aware of large birds, so he was not inclined to investigate the cave either.

In the summer of 2009 I took the train to Pukatawagan again with the express purpose of checking out the cave I just mentioned. Not only were Gabe and Adlerd aware of the mystery surrounding the area, but so was Angus, my host from Puk. The cave was alleged to be within a mile or so of where he had challenged that nocturnal avian visitor many years earlier.

After finding a small boat and motor that could be carried over the half-mile portage, and a crew to help with the task, three of us boated to the mountains that we hoped would offer up the cave without too much difficulty. We were wrong. Hours later, after going much further than half the distance of my endurance, my legs gave unequivocal indication that I would *not* be able to retrace my steps over several mountains divided by excruciating deadfall. I faced the very real possibility of spending the night in the mosquito-infested bush unless an easier route was found back to our boat. Fortunately, and providentially, after taking a route that seemed to be the opposite of the planned direction, we came upon a creek that we assumed might lead back to our starting-point. It did indeed, and we made it home in good time, not minding at all that our prowess in the woods could not account for our safe return.

Some months later, after the memories had faded, and the healing and licking of wounds was past, the desire for discovery beckoned again, so a trip to a different cave was planned. This one—the original cave—was so remote that a float plane was the only feasible transportation. Fortunately, a little lake provided access, and an X on the map of the mountain beside it assured us of success.

Ken Reader, who had been interested in my discoveries for some time and had accompanied me on several previous excursions, was delighted to camp on that isolated shore and scour the mountains for signs of the cave. He led me around expertly with the use of his GPS, all the while suffering from a leg injury which inhibited him considerably, but which allowed me to keep up comfortably. We covered what we thought was the appropriate territory with no sign or smell of the cave, and returned home with our accustomed disappointment, but also with fond memories of the experience. Knowing that we must have been close to the cave site, I sent a map back to Miles showing where we had walked in relation to his X, and soon I learned that his mark was, in fact, on the wrong side of the lake, but within a mile of where we had searched. So now 2010 became the year of hope.

2010 came and went, with no practical opportunities to investigate further. Then, in the winter of 2011, conditions seemed right to tackle the search by winter road

Sketch of two huge bat-like creatures made by one of two witnesses who saw them gliding overhead just before dark in the summer of 2012 in southern Manitoba.

and snowmobiles, but success eluded us. A host of difficulties hampered our efforts, so summer again seemed to be a more practical time to explore.

Thus the mysteries continue. However, with each new piece of information that surfaces, the truth is coming ever nearer, and efforts to locate some of the caves will hopefully soon reveal interesting details.

Two men saw a pair of big birds fly over them some years ago in an area north of South Indian Lake. The one I spoke with said that the bodies were the size of humans, and their yellow claws were bigger than a man's hand. The big beaks were also yellow.

At the northwest tip of Manitoba's Interlake region, in the summer of 2003, a giant bald eagle was spotted sitting in a tree beside a highway. The passenger who saw it had difficulty persuading the driver to believe him, but, after gassing up nearby, they returned to the site. The passenger, convinced that the appearance of this bird was a rare spiritual phenomenon, borrowed a cigarette from the driver and gingerly crossed the ditch in order to offer it some tobacco. It stared steadily at him, and when he got close, it flapped its wings forward in a fanning motion towards him. The wingspan was "at least 16 feet," and the height of it was over five feet. The body was wider than a man's, and the head, identical to the white head of a bald eagle, was well over a foot wide.

There are rumors of other such sightings in the area, although few birds appear to be as big as this one.

Not too far away, some conservation officers were parked in a hayfield near the northern tip of Lake Manitoba one night in order to monitor illegal activity. It was pitch dark, but they knew that a flock of geese was just a few feet away. Suddenly there was the sound of a big bird descending, and then the squawking of a goose as it was taken up and away.

A man from Fairford recalled his dad seeing a big black bird around 1950 that had an incredible wingspan, judging from the width of the clearing where it flew up. Another Fairford man, while out on a nearby lake, thought he saw an airplane approaching, but it turned out to be a big bird.

In southwestern Manitoba, a family out driving one spring saw a huge flock of eagles in a hayfield. One in particular stood several feet higher than a round hay bale nearby, making the others look "like crows." In the same general area, a man had seen a big "eagle" eating a deer, and its size had made the deer look like a rabbit in comparison.

Several people remarked about the difficulty the big birds seemed to have in getting airborne, taking many running steps in order to do so. Perhaps some species do not have the same ability to achieve flight as readily as others, or they may not have been mature enough to take off effortlessly.

A firefighter from Puk, after someone suggested I look him up, told me that he had seen some unusual birds when he had been on a helicopter patrol flight some years ago. He had been looking down at a river several thousand feet below when he had spotted three large white birds gliding just above the water, "the size of small airplanes." He notified the pilot, but in the time it took to turn back, there was nothing more to be seen. When I enquired about the nature of the riverbanks, he indicated that they consisted of high, rocky formations, so we both agreed that the birds could easily have arrived at their lofty nesting sites or cave during the time that he lost sight of them.

Perhaps these are the same species of large feathered birds that two ladies saw flying over The Pas one overcast summer day. They thought it was an airplane, but since it had no numbers under its wings, and was making no noise, they concluded that it had to be a big bird. It did have some black markings on its extremities, something which the firefighter from Puk was probably not able to discern because of the distance.

A tip from my son put me in touch with a young man from Norway House. Not many years ago, he was nearing his community one night after a long ride home from Winnipeg, when the driver suddenly pointed out a large bird in front of them which was flying at eye level in the same direction they were driving. Its wings spanned the whole road, and its tail's width matched the size of the car's hood. It had flapped its wings about four times before it got away from them, and each time the tail had lifted, exposing claws larger than a man's hand. Its shape and color resembled a golden eagle.

Recalling the golden eagle that flew off with the deer, this was perhaps an even larger version of that species.

Although this appears to make three separate categories of large, flying, bird-like creatures, there may very well be more. I am still puzzled about the ones that some think are responsible for eating bark off poplar trees. From the description of the distance between the place where the claws appeared to grasp the tree and the area above it that is scraped clean, it would appear that this bird might only be a few times bigger than the large pileated woodpecker. However, to date, I have not heard of such a one, and as to where they might live, that also is a mystery. The builders of the big nests I described earlier I presume to be the huge feathered species that different folks have seen in daylight. They do not seem to be nocturnal. The huge "bats" on the other hand, seem to be strictly nocturnal, and, like their smaller relatives, possibly inhabit the caves whose darkness they would prefer in daylight hours, and whose environs show every evidence of their presence. Their size seems to be similar to that of the larger feathered birds, as both have been compared to small airplanes.

Tracks reported by a few hunters. Could they be made by large birds?

And, the mysteries increase! In July 2009 a couple from the Opaskwayak Cree Nation at The Pas saw flying creatures that they had never seen or even heard of before. Husband and wife each witnessed a lone specimen separately, about a week apart, and I happened to talk to each one separately as well, several months apart. It seems that just at dusk, as loud fireworks were beginning on Canada Day, Dana saw a large bat-like creature emerge from some trees and fly across the yard. She described the wings as bat-like, but the body shape reminded her of a de-feathered goose. Her husband, Ernie, saw it one evening too, but still in broad daylight. Its body, as it had hovered for a while in an upright position, appeared "like a little man" because of the extended arms that were attached to the wings, the hands that seemed to have individual fingers, and the legs which were spread in flight.

A short-haired, medium-sized dog was in the house with us, so Ernie lifted the front of it up in the air, indicating that the large rib cage (turkey breast) and splayed hind legs made it a close match. The wingspan was about seven feet, with the total weight of the creature somewhere in the 40-pound range. His sketch showed a bat-like wing shape, just like his wife described. The color was brownish-black, with the short hair being on the front, and longer fur on its back. Its size and ability to hover sound similar to some flying foxes.

In the Yukon, some villagers claimed to see a bird the size of a Volkswagon flying off with a rabbit in its beak. It was pointed out that eagles, however, typically carry prey in their claws.

A story out of Alaska alleged their bird to be larger than an Otter airplane.

Some of the sightings I have related pertain to birds that are known and easily recognized, like the bald eagle and the golden eagle, only in an enlarged form. Others are unique and have no precedent for comparison, simply appearing to be a new species of feathered monster bird, some brown, some black, and some white with black markings. Then there are the huge "bats" which do not appear to have feathers, but rather seem to have a smooth skin-like covering.

As more stories come in, and as investigations continue, we will no doubt soon be able to build a credible body of information about these remarkable creatures.

Rock paintings depicting big birds have also been reported in the north, which if correct, would indicate that earlier peoples knew more about them than we do today.

Although the following account does not fit into the category of unknown birds, it does describe unusual bird behavior, and I feel it is worthy of mention. My friend and host from Pukatawagan, Angus Linklater, heard the story from the father himself. A husband and his wife were out hunting, and they had in the canoe their baby, wrapped up in typical Native fashion. The man had shot a duck, and both parents had left the baby behind in the canoe while they went to retrieve the bird. When they returned, the child was gone, and the distraught parents looked in the water and all around, desperately hoping to find the missing child. It was then that they heard its cries from somewhere up high, and noticed an eagle's nest in a tree. The father grabbed some rope and quickly climbed the tree, which fortunately had plenty of branches. They had made a great deal of noise to distract the birds from their prey, so that when the man reached the nest, the baby was unharmed. He lowered it to the ground with the rope, into the arms of the grateful mother.

With this and other stories in mind, it comes as no surprise to hear that parents sometimes made their small children hide in the bottom of canoes when crossing open water in certain places.

BEAVER DUCKS

THE DUCK-BILLED PLATYPUS of Australia—that egg-laying, milk-producing anomaly that we know so well because of its unique characteristics—may just have a distant relative tucked away here and there in our vast North American landscape.

Strange as that may seem, there have been a number of sightings brought to my attention.

The very first mention of the word "platypus" was heard in the northern community of Nelson House. It was a head-shaker for sure, but I recorded the information anyway, and resolved to check it out sometime . . . no hurry. But when another party recommended that I go see so-and-so about a platypus sighting, my curiosity drove me to the residence of a young couple who gave me their account.

In the mid 1990s, husband and wife were setting out one morning from the shore of a lake not far from their community when they heard a splash behind them. Thinking that it was probably a beaver, they paddled on. Shortly after, they both saw what looked like a beaver swim under their canoe, only it had a wide, flat bill. The creature's feet and beak-edges had an orangey hue, but other than that, it appeared similar in size and color to a beaver.

In the Lake Manitoba First Nation community in southern Manitoba, a young man, after hearing this story, said he shot at one as a boy. Its fur was blackish-brown like a beaver's, but he felt it was slightly larger than one. The feet were webbed, too, but bigger than a beaver's. Several others in the community, including his grandfather, recalled hearing elders speak of the creature.

At Moose Lake, near The Pas, Manitoba, Leonard Nasikapow was driving down the road in daylight when he had to stop for a beaver-sized creature that came out from the bush towards his truck. It was brown like a beaver, had short legs, and ran in a fashion similar to one. Its six- to seven-inch-long duck-like bill was grayish-black, and the seven- to eight-inch-long tail had fur on it. He never stopped to examine it further, assuming that the animal lived in the area.

Jarmo Sinisalo's rendition based on several descriptions and sketches of the Manitoba platypus.

Others in the community had heard of such a creature before, but knowledge of it was not widespread.

Another man I spoke with in a different community knew of a trapper in southern Manitoba who claimed to have caught a beaver once that had a beak.

Some folks on the Shamattawa First Nation were also familiar with this "duck-billed beaver," but nothing specific. Northwestern Ontario was also mentioned with respect to sightings.

At Pukatawagan, an isolated reservation north of Flin Flon, there was limited knowledge about the creature. Like in other communities, most of what was known was passed down by the elders. However, Susan Lackie, a teacher who had lived in that community in 1989, said she and those with her in a boat were surprised one evening to see something odd hurry off the shore of an island as they approached. It resembled a beaver, but had a large, dark beak with orangey sides.

Stories of this animal also persist around Reindeer Lake, a large body of water found on both sides of the Saskatchewan-Manitoba border in the north.

Near Lake of the Woods, on the Ontario side, is where a young boy encountered a big-billed creature in a ditch on his way to school around 1950. He killed it by stoning, and showed it to his parents. They had no idea what it was, so he took it to school. His teacher found a picture of one in a book, so, although they identified the animal as being a platypus, it just got thrown away.

Now, almost 60 years later, Les Nelson described it to me as being 12-14 inches long, not as big around as a beaver, with a bill that was much bigger than that of a duck. Being of such a small size compared to other sightings, this one may just have been a juvenile.

A man trapping in eastern Saskatchewan not far from the Saskatchewan River told his friend in The Pas, Manitoba, that he had caught a funny beaver in a beaver lodge. It can therefore be assumed that perhaps these creatures go in and out of beaver houses on occasion.

Folks from Split Lake, east of Thompson, also mentioned sightings of them in the Nelson River.

The various communities that have some knowledge of this animal have different names to describe it, but the bill seems always to be the outstanding feature. The most common name, when translated into English, is "beaver duck" or "duck beaver," but I heard "duck mole" mentioned too.

Based on the locations of the various sightings to date, it appears that its range can be broadly described as the central portion of Central Canada.

CROCODILES AND LIZARDS

WE KNOW OF SALAMANDERS AND LIZARDS in the southern parts of Canada, alligators and crocodiles in the southeastern United States, but who ever heard of crocodilians in the regions of Canada where lake ice gets thick enough in winter to support loaded semis as they haul freight to remote, roadless communities? Nonetheless, I hear of them in my travels.

My very first such report came from the north end of Lake Winnipeg, near Norway House, where a trapper had told of chopping a six foot crocodile or lizard out of the ice in a creek before World War One.

Next, a man and his sister from Nelson House recounted their experience as children when the family took one of their rare long trips in summer by canoe to the area of their winter trapline. Somewhere along the way, in their wanderings, they came upon some holes with little crocodilians crawling around. Inside the "nests," which were simply holes about a foot in diameter, and about a foot and a half deep in the grassy bank near the water, were the open eggs as well as some which had not yet hatched. These were about six inches in length, and about four inches in diameter. Apparently the siblings did not linger in the area, perhaps because of the same question we might ponder—how big is the mother?

I have heard a few similar reports from that region, including crocodile sightings, generally by elders of a previous generation.

One man from Pukatawagan whom I spoke with killed a brown, four-foot crocodile-like creature in a log trap. It scared him so badly that he never went back to that spot.

A number of encounters circulate in Nelson House where people traveling by canoe, especially in lakes to the north, were attacked—unless perhaps it was their paddles being attacked. One man is alleged to have killed a small crocodilian with his paddle when it became aggressive.

Another man is alleged to have shot a duck in a marsh, whereupon a crocodile surfaced and grabbed it.

Newly-hatched lizard seen by young
François siblings north of Nelson House.

An old lady from South Indian Lake tells an intriguing story about a creature which seems to fit the description of a crocodile. When she was just a young girl, her father had been out on the land, camped beside the water of a creek. At dusk, he heard and then saw a creature come out of the bush towards the water, its eyes like glowing lights. He killed it with six shots from his .250 rifle. He had been known as a fearless hunter, but this creature apparently shook him up so much that he recruited his brother-in-law the next day to help him burn it "so that it would never come alive again." It had taken a good deal of wood a long time to burn it up.

Several days later when her dad brought the family to the spot, the water was still bloody. This animal was described as being around 15 feet long, green in color, having a long, pointed mouth with lots of teeth, big eyes, a long tail, and hard bumpy scales. It was called a "bad animal" in the Cree language. The storyteller, who was the hunter's daughter, referred to the creature as an alligator.

Moving into south-central Manitoba, where the appearance of these creatures could be considered somewhat more likely, reports from residents along the east shore of Lake Winnipeg mention big crocodile-like animals sunning themselves, or just resting on flat rocks beside the water.

One lady who lived beside the lake at Bloodvein saw what appeared to be a large crocodile on flat rocks near her house. Her son told me that she sold her home because of it, and the new owner soon saw the same thing. When he went outside early one morning, he noticed something unusual on a flat area about 150 feet away. A large dark grey crocodilian whirled around on its legs and went into the water. Other reports include tracks of a big creature that "drags its belly."

At Berens River, a man, fishing at the rapids, felt he was being watched, so he looked down into the water, and there, below him, was the creature shown in the sketch.

In the same lake, except further to the south where the Winnipeg River enters Lake Winnipeg, a teen had been walking across the power dam when he decided to look over the side into the water. There, a few feet below the surface, was a larger version of the "mud puppies" they would sometimes snag with their fishhooks in the river—a small lizard-like black salamander. This one, with its large bulbous head, was identical in all aspects except in size, for it was about ten feet long. It swam like a crocodile using its legs, but its movement also had a snake-like appearance.

This raises the question about continuous growth in reptiles and amphibians—and sightings such as this one would seem to confirm that possibility in the extreme.

When I was told that a couple from Little Black River (a community also on the east shore of Lake Winnipeg) had seen something larger than an otter on the ice one spring, I expected a description of a bear-sized beaver since that is one of their habits. But when the words "Komodo dragon" were used, it took me by surprise until I realized that the majority of lizards and crocodilians that had been described to me were in or near Lake Winnipeg. This particular creature was seen running on the ice, diving into the open water for a while, and then climbing back onto the ice to run some more. Its legs were on the sides of a long, low body, estimated at five feet, and from a distance it looked much like a Komodo dragon. The rumor of an "iguana" seen on a nearby beach may have some relation to this scenario as well.

At Poplar River a man saw a three-foot-long lizard in the grass, its body about nine inches in diameter. He mentioned that he had heard of others who saw something like it "half flying on top of the water."

On a reservation near the Lake Manitoba Narrows, crocodiles were reportedly seen in certain creeks in times past.

In the 1950s, several brothers saw "an alligator with a jagged back sunning itself beside a lake" in central Alberta. In northern Saskatchewan, a creature with a back resembling a crocodile's was seen.

A teacher, C. P. Alarie, living one summer in a district bordering Lake Manitoba north of Toutes Aides gives the following account in a letter:

> The marine monster report from Sandy Point has greatly interested me. I am very happy to learn that investigation and study of the mystery is on the way.
>
> I was indeed the first teacher in that forlorn, abandoned district for three summer months in the year 1948. Everybody was talking about the so-called "Big Snake." I didn't believe a thing of it at first, relating their bewilderment either to local gossip or superstition. But, one Sunday, as I went to Steep Rock Lake with my faithful good-hearted limping friend Marcien Fleury (nick-named Solemn), I indeed saw a huge brown thing, jumped two feet and then stayed still.

This crocodilian was seen and sketched by Bradley Flett of Berens River.

This skull was found and sketched by Darrell Moore of Nelson House.

> The size of an alligator. I personally thought it was a fierce sort of animal and admit I was dreadfully scared-never so afraid in all my life. At about a small distance from where it was I saw the entrance of its cave. I then wrote to the Department of Research in Winnipeg and never had any answer whatsoever. . . .
>
> I hope further research will soon be underway and still think that the possible Ogopogo could be brought to a zoo either at National Park or Winnipeg.

This gentleman obviously saw something different from the "big snake" that everybody was talking about, and since he described it as something looking like an alligator, then that may have been what it was. After all, this was the same body of water where such sightings were mentioned near Dog Lake and Lake Manitoba Narrows to the south.

A rather unusual story is told of a grandpa's experience in the isolated northwestern Manitoba wilderness one fall. He had discovered a hole in a sandy bank, and on investigating it, was chased by the occupant. It was described as a crocodile-like creature, but it seemed to give off light when it moved. He had only been able to kill it by shooting into a softer part of its skin, which was behind a leg. When it fell, its heavy body had cracked the ice. Its skin had felt and sounded like iron scales when hit with an axe.

In the same general area, but in recent years, a trucker traveling the highway between Nelson House and Leaf Rapids saw what resembled a crocodile partly submerged in the water near the road. One end of it had a large, bulbous feature, much larger than the snout of an ordinary alligator.

At Oxford House a man saw a gecko-like creature on the ground among the trees in his back yard. It was about two feet long from nose to end of tail, and it moved very quickly on its long thin legs.

An outfitter from God's Lake Narrows reported seeing a similar, but smaller version of it running on top of the water near shore!

On the shore of a lake near Nelson House was a skull, a few years ago, of something that strongly resembled that of a crocodile. It had large eye cavities on the top of its foot-wide head, and the row of teeth was about three feet long.

Putting all the accounts together would suggest that perhaps we do have crocodilians in Canada, and the rumors of pictographs in the north depicting them add even more support to that possibility.

UNDERWATER MOOSE

OF ALL THE UNUSUAL CREATURES I have been told about, this has to be one of the weirdest.

Why? Because it just doesn't make any sense at all.

Sure, a moose can keep its head under water while it's feeding, for at least five minutes, but that's a far cry from *living* under water. After all, moose are not amphibians. They need to come up for air!

Not so with these critters. But perhaps they're not moose at all.

Only thing is, they look like them. Well, not exactly. But close enough that the Native people call them moose. "Etampeegomoosowa," which in Cree means "underwater moose"; or "manamoosa," another name I heard.

There is just no other creature in the north that bears a closer resemblance. And yet, if the two species were standing side by side, they might not look any more similar than a horse and a cow would.

Let's compare the two "moose." Or perhaps contrasting them would be more appropriate. In size, the underwater moose is stockier—a heavy body, but with much shorter legs. And the stomach sags way down. It has cloven hooves, like the regular moose, but with an abundance of hair around them. The antler descriptions varied. Some have been seen with little sticks for antlers, while others were at least as big as ordinary moose horns, but "branchier." Some had none at all. One man noted that the antlers on the one he saw seemed to grow upwards from the head rather than sideways.

The one feature, and perhaps the only one that resembles a moose, is the head, and even so, a number of witnesses noted that the head seemed smaller than that of an ordinary moose.

Don't ask me why an underwater creature needs antlers. But who knows, maybe for the same reason that land animals do. As for color, it is brown on top, and the rest is grey, whereas the ordinary moose is mostly dark brown to black.

STRANGE CREATURES SELDOM SEEN

A feature that was mentioned in the majority of reports is the sand that is visible in the hair. This would appear to be an obvious result of lying on the bottom of lakes and rivers where they were occasionally spotted.

Let's go to the underwater moose capital of Canada—Pukatawagan, Manitoba. This is where I first heard of the creature in the winter of 2003-2004 after driving for hours, north of Flin Flon, on the winter road over a half dozen delightful lakes. Lakes which in the summer release all manner of wildlife, including the underwater moose.

The following are some of the incredible stories I heard.

Two men went moose hunting in Morin Lake (the second-last lake that makes up the winter road as you travel north), and built a fire after dark beside their camp on the shore. They called for moose, and then, when nothing seemed to be happening, they turned in for the night. But suddenly there was the sound of something coming out of the water. The narrator of this story, Paul, grabbed his flashlight and shone it on a "moose," noticing a yellow reflection from its eyes. Before the animal was fully out of the water, his partner began shooting, emptying his .303 to no avail, even though it was very close range. Then Paul emptied his .270 on it with the same results. Each shot, he said, sounded like a bullet hitting an empty bag.

The large rack was most desirable for winning the fall moose antler competition, so they decided to go after it in their canoe. As they caught up with it, Paul reloaded his gun and asked his partner to shine the light on it, hoping that it wouldn't turn on them and flip their canoe. But, as soon as the light appeared, the beast sank down into the water and disappeared.

Paul hung around for three days—the length of time a dead animal would normally take to rise to the surface—but nothing happened. Then he came back a week later to check, but still nothing. He knew then the truth of a story from an elder who had experienced the same thing in that very lake, wasting about a dozen shells.

Another man from Pukatawagan told me about his encounter with this unusual creature—he recalled it was probably in 2003. He and his partner had been out moose hunting beside a lake when they spotted antlers above the water as a creature of some kind was swimming a distance of about half a mile to the opposite shore. When it landed and went into the woods, the man and his partner tried to bring it back with the typical moose calls, and succeeded. Again, only its rack was visible as it swam towards them, and when it emerged from the water, one of the men got away one good shot with a .303. The moose "fell" back into the water and remained there out of sight. The men set about making some tea, and were surprised almost an hour later when the creature emerged from the water and ran into the bush as if it had not been wounded at all. They followed it, but could not catch up with it. In describing the position of the antlers in comparison with an ordinary moose, the hunter raised his arms straight above his head.

Also near Puk, in the middle of Morin Lake, two hunters came upon a couple of antlered moose in the water, and got within 20 feet of them. Both had .30-30's, so one fired six shots at one, and the partner seven shots at the other. Neither one killed his animal! One of the men, who related the story, stated that one shot to a normal moose would have definitely brought it down at that close range. He went on to say that he fired at another such creature later the same summer in the same lake, at even closer range, and his four shots wouldn't penetrate. He did remark, however, that he thought they looked different from ordinary moose because of a smaller head and the grey color.

I heard of a man in a different community who hadn't just emptied his clip on the creature, but an entire box of shells. And all that after being given advice from another's voice of experience that he shouldn't bother at all.

It is common knowledge that the hide of this creature is very thick and tough—and sandy, almost impenetrable. One group of hunters saw their bullets draw blood as the animal ran along the shore, but they were not able to bring it down.

Sketch of "underwater moose" by Paul Dumas of Pukatawagan showing one actual and one conventional moose antler.

One of the few men who ever managed to kill one shot his in the eye. The meat was a yellowish color, and the layer of fat very thick.

One lady who still remembers seeing one that her father shot in Granville Lake said that the meat was green. He had been able to bring it down by shooting it in the ear, but because the meat was not tasty, they left it behind.

I hear accounts from hunters of different communities who saw what they thought were moose tracks going into the water, but not coming back out, and when they checked the next day, there was still nothing. Then when they paddled out into the lake and looked down, they saw big forms lying on the bottom!

A man from South Indian Lake tells that when his dad shot at an underwater moose, his mother tried to stop him because she thought it was a horse. And in the same vein—a man marooned on Berens Island in Lake Winnipeg reported seeing two horses running towards him along the shore until they noticed him—and then veered off into the water and disappeared. In the same area, a man saw a deer-like animal jump into the lake from a small, rocky island and swim away. His nephew told me that his uncle circled the island looking for it, but it just vanished.

A lady from Puk thought they were cows she saw running along the shore, but they also ran into the water and vanished.

A different Paul, who grew up at South Indian, told me that when he was about 14 his family had been out on the land somewhere between MacKerracher Lake and Denison Lake northwest of South Indian. One day when his father was away beaver hunting, they saw an underwater moose. His mother fired six shots at it with a .30-30, but it just kept on running. He said it had short legs and ran with the jerky gait of a pig. He also observed that its fur was full of sand.

Another man compared the gait of the creature he saw in a ditch south of Grand Rapids to that of a bear, with its short-legged choppy movement. When he got closer to it, he saw that it was moose-like, and had antlers, but had a low-slung belly.

A hunter and his wife from Nelson House encountered this creature on the ice near the end of winter not many years ago. He emptied his gun on it again and again, hitting it with every shot since it was only walking. He remembered it being brown on top, and grey below, with the typical short legs. His grandfather had also seen moose lying at the bottom of the lake.

This next story may help us understand why these creatures are occasionally seen on land or ice.

A former conservation officer from the community of Nelson House related an experience that his grandfather had told about—the result of one of his winter trips by dog team across the north. A "moose," he thought, was lying down in the snow near shore, so he managed to kill it. He soon found out why it did not get up or run away. When he rolled it over, he discovered that it had been standing up all along, but because its legs were so short, its stomach so low, and the snow so deep, that it just couldn't move. Perhaps it was also very weak.

He remarked on the small size of its head in proportion to the large body. Also, it had big, wide pads around its hooves that were covered with hair, giving it something of a webbed appearance. On skinning it, he found that it had an extremely thick layer of fat, and the skin around its neck was a series of flaps covered with long hair. He speculated that this feature, not present on a typical moose, might be an apparatus that had to do with obtaining oxygen under water.

The hunter came to some conclusions after pondering the situation. He noticed the ice broken up in the area, so he believed that the creature had gone in and out of the lake in order to forage on the willows growing nearby until just a few days ago. Then one of two scenarios left it stranded above the ice. Either a very cold snap prevented it from being able to break the ice in order to get back into the lake, or a very heavy snowfall made it impossible for the creature to move—or both. Since it was immobilized, it was easy for the hunter to shoot the animal in a vulnerable spot with his .30-30.

An intriguing recent development sheds some new light on the above account, and perhaps provides the needed evidence required to substantiate information already gleaned about the enigmatic underwater moose.

In the fall of 2009, commercial fishers were engaged in their trade in different lakes, setting and lifting nets. To the north of Nelson House, in the Mynarski Lakes, the Spence family moved to their camp, and fished nearby. One boat was operated by Conrad, his wife Veronica, and helper Henry Wood. On this particular day they were lifting nets at a narrow part of West Mynarski Lake, and this is where the story begins.

Veronica heard a noise nearby that sounded like an animal gasping for air, and seeing a set of antlers on the head of a creature that had just broken the surface, she drew it to the attention of her two partners. Not only had the animal not come swimming from shore as would have been expected, but it appeared suddenly in the middle of the lake—wearing the strange antlers that you see in the picture.

Not even knowing for sure what kind of animal it was, their thoughts of fresh meat caused them to formulate a plan very quickly, especially since they had forgotten to bring their store-bought meat with them on this trip. Conrad took over the controls and circled the animal, cutting it off from swimming towards shore, which he assumed it would do. Since its antlers afforded an ideal place to tie rope, Henry was dispatched to do the dangerous job of tying their small net line to the antlers—the sharp points not only endangered the aluminum boat, but also anyone who came near them. When he grabbed an antler, the animal almost pulled him overboard, shaking him up enough that Conrad relinquished control of the boat to Veronica so he could help with the delicate task of securing the animal.

When they finally managed to tie it to the front of the boat, Veronica put the motor in reverse—only to have the small rope snap. She drove forward, and the rope was re-attached. Three times the rope broke when she backed up the boat, and one time when she drove forward too far, the boat was even on top of the animal. Almost they gave up—but the fourth attempt was successful, perhaps because they tied the rope to the back of the boat. Henry's jacket shrouded the creature's head, which was now under water. Drowning a moose by keeping its head under water was an age-old method of acquiring meat when a gun or axe was not available in the canoe or boat. The difference here, however, was that this "moose" didn't co-operate and drown in a minute or two as did normal moose, but instead took much longer. Perhaps the only reason why it eventually drowned at all was because they pulled it steadily behind the boat with its head underwater, forcing more water through its system than normal.

They towed it to the nearby shore and took off for camp to recruit help and supplies for the job of skinning the animal and cutting up the meat. Fourteen sets of hands were soon employed in the attempt to remove the beast from the water—13 pulling on four ropes tied to different parts of the animal, and Conrad pushing from in the water. Normally two or three people could pull a moose ashore, but this seemed not to be an ordinary one. Not only did it have the weird antlers, but it had short legs, a low-slung belly on an unusually stocky body, large feet with long, overlapping hair around its ankles, a smaller head than the normal moose, longer, more pointed ears facing back, a wide nose, and the odd colors of brown on its back and black the rest of the way down. Someone also recalled seeing sand in its hair. Then there was this "scarf" under its head, running the length of its lower jaw—reminiscent of the "flaps" mentioned in the earlier account. As you can see from the picture (p. 179), the "scarf" or "folds" under the chin are clearly visible. This creature, living beneath the surface of the water most of its life, may have some mechanism that either enables it to acquire oxygen or to conserve it for long periods of time. This unusual feature definitely requires further investigation.

Everyone I talked to who had participated in the dressing procedures remarked on the layer of fat. Thickness estimates ranged from four to six inches. The fat was described as being more yellowish than the usual white, with even a purplish tinge.

The meat had been good. Conrad and Veronica insisted that the meat didn't have the flavor of typical moose meat, but was rich and tasty—and eaten three times a day until it was all gone.

Veronica and Conrad Spence holding the antlers of an 'underwater moose.'

Conrad's father and grandfather were present also, and the older man was heard to say that it had been more than 50 years since such an animal had been killed.

I was delighted, of course, to hear this account, thinking that the antlers (which were the only part of the animal saved) would obviously be proof of the existence of the unique underwater moose. You can imagine my disappointment when the DNA results came back simply as "moose." The fact remains, however, that this animal, together with so many others that match its description, share unique traits that, although they defy explanation at this point, indicate a type of "moose" that is significantly different from the ordinary.

(Some more recent information is in order here, however. Since I had some DNA work done by an acquaintance from Louisiana years ago, I decided to try him for a second opinion—and I am glad I did. His genetic comparisons indicated a closer link to the European and Asian moose than to the North American. Another DNA specialist suggested that these creatures were possibly hybrids.)

"How can you drown a moose that lives practically its entire life underwater?" is the question I asked Conrad. His cogent answer, as I recall, was that even if it somehow gets oxygen from the water, when too much was forced through its mouth, it still drowned.

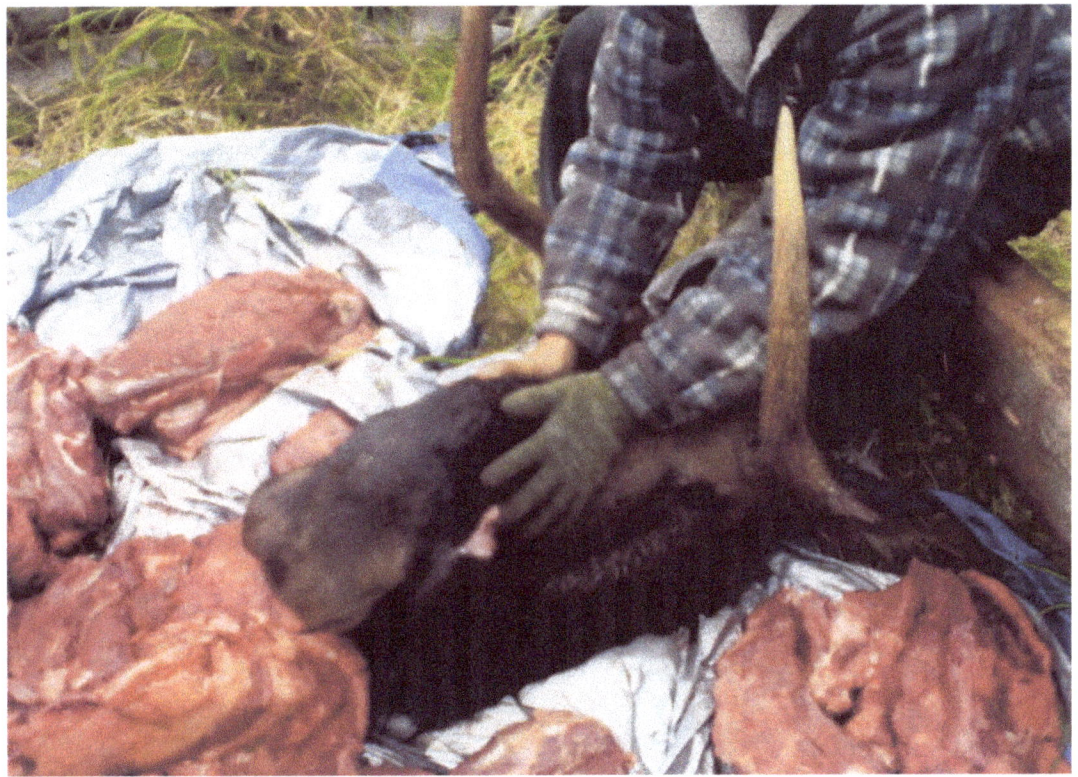

Side view of the "underwater moose" head.
Notice the black "scarf", or "folds" under the chin.

Another question that provokes the mind is why the creature didn't try harder to escape when it had many chances to do so. A typical moose has no choice but to swim with its head out of the water, making it extremely vulnerable, but this one, and others I've heard of, appear not always to exercise their ability to submerge in order to escape. Perhaps once they decide to rise above the water and begin using their lungs, they are unable or unwilling to revert to their underwater system quickly. And yet there are numerous accounts of these creatures running into the water and disappearing. Fortunately, Veronica had her camera along—even if it only captured the head and antlers—and the "scarf" under its head—which is so plain to see.

The only characteristic that didn't quite match most of the other accounts is its black coloration, where I expected grey, so, considering color to be a variable and minor feature, I feel confident that those three fishers from Nelson House did indeed bag an underwater moose, and fortunately survived to tell about it.

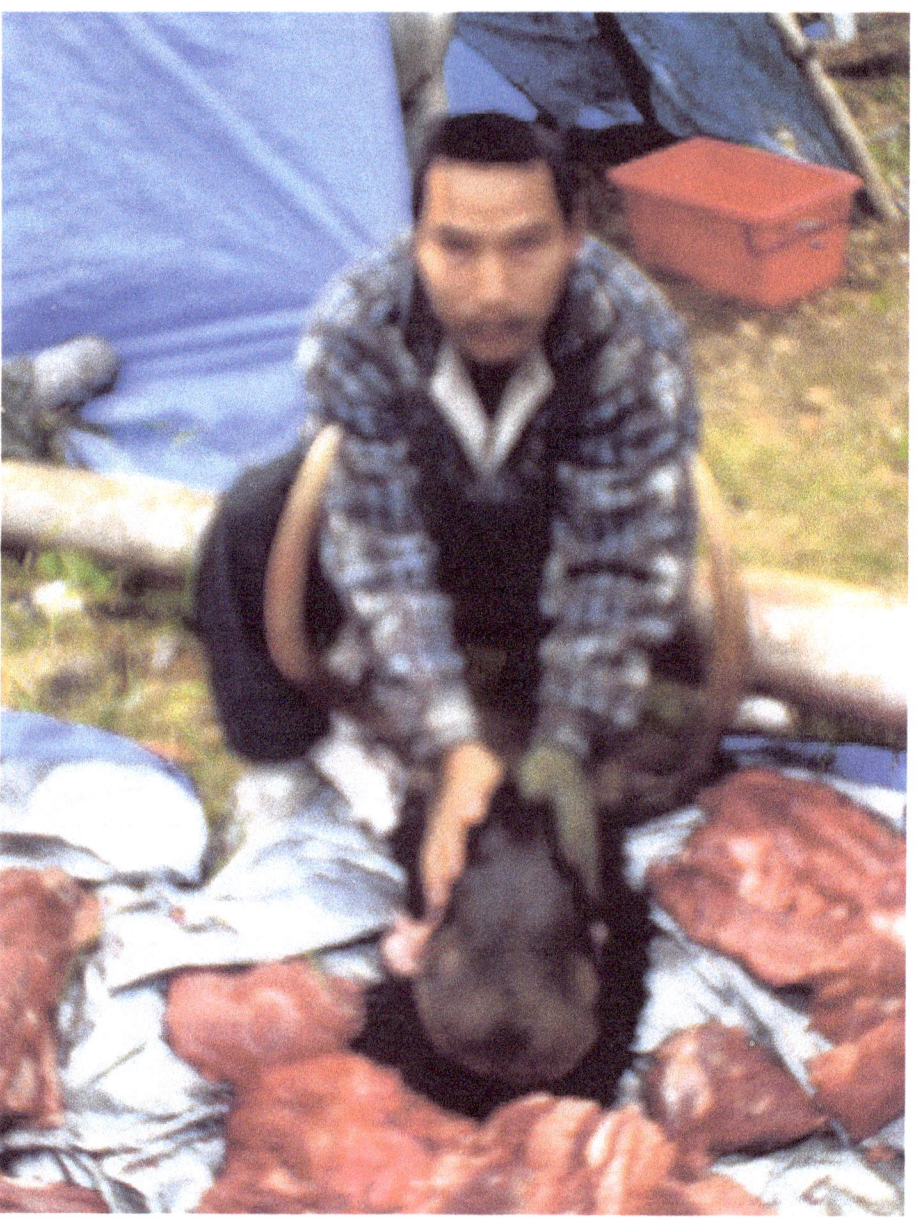

Henry Wood posing with the head of the unusual "moose."

A Moose Lake resident recalls his father's account of shooting an underwater moose. It had gotten away from him by going into the water, but the next day he managed to shoot it. He remarked about its short legs, the more than two inches of fat, its horns, and that it was good eating.

Someone commented that this creature, or perhaps another similar to it, had a horse-like tail.

A man from Poplar River, a community on the east shore of Lake Winnipeg, recalled hearing that someone had seen a rotting animal on an island near McBeth Point that was the size of a cow, but with short legs.

Out of Tetlin, Alaska, near the Yukon border, comes a story akin to the ones already shared. Three sisters told how their dad had seen what he thought was a moose's horn under the water. He had rolled up his sleeve, leaned over the edge of his canoe, and proceeded to pull on the dead animal. Apparently it wasn't dead, and, as it rose, he had to scramble to make it ashore, stepping on its back in the process. The ordeal left him so frightened that he carried his canoe all the way home instead of letting it carry him.

Near where I live, at the northwest tip of Manitoba's Interlake, a man I have known for some time told me about a strange "moose" sighting his dad reported in a small local lake. Its head had looked odd to him somehow, and the creature had not raised it for chewing as regularly as expected. He wondered if it was perhaps an injured moose, but it eventually disappeared under the water.

Perhaps this is the creature that has, on occasion, been referred to as a water buffalo, and perhaps that name will prove to be more accurate than any other.

Wouldn't it be nice if a specimen, or a photograph of one, became available? Then we could finally determine what this elusive, water-loving animal is all about.

SEA DOGS

IT MAY NOT BE A CANINE AT ALL, but, as in the case of the underwater moose, because this creature appears to be more dog-like than anything else, it is referred to as a dog that lives in the water.

Except for the two vivid eyewitness accounts that were shared with me in 2008, other stories have been few in number, and sadly lacking in details, simply because the creatures were so extremely elusive, and possibly nocturnal as well.

I had assumed for years that the animal purported to have stolen fish out of a boat was one of the various Manipogos of Lake Manitoba, but when I finally, many years later, discovered that the witness was still available, I found a remarkably unique and shocking story, one which finally brought some satisfactory information to bear regarding what may be this mystery dog.

I had heard tales before of dog-like creatures that came out of the water, left their tracks behind, and returned whence they came rather quickly. I was even given the impression that Dog Creek and Dog Lake, both near the Lake Manitoba Narrows, were named after these elusive creatures, since people apparently had seen them come from and return to these waters. The man, Mayt, who sighted one of the Manipogos, told me that his aunt insisted on numerous occasions that she had seen a dog similar to a Dalmatian disappear into Dog Creek when it was approached. In the north, some communities had similar vague references to Sea Dogs, with few good sightings to substantiate them.

I paid Mrs. Kaartinen a visit in the winter of 2009 after several telephone conversations the previous fall, and came away with wonderful information on what I believe is Manitoba's water dog.

The incident in question took place one summer in the 1980s after Mrs. Kaartinen's sons left some junk fish in a boat at the edge of the lake. She walked the short distance from her yard to the boat to get a pailfull of the fish to cook up for the chickens, and when she got there, a creature that "almost filled the back of the boat" whirled around and jumped out the back into the water. It swam away with its head still visible, so she followed it for a while as it swam parallel to the shore.

She described it as being bigger and longer than a dog, with a larger-than-human head extending forward from the body instead of rising above it on an arched neck like a typical dog. Its muzzle was shorter than a dog's, with some "pouch-like bulges" beside it (which might have appeared so if its mouth were full of fish).

Mrs. Kaartinen noticed that its legs were shorter than those of a typical farm dog, and there was a bit of tail at the end of the long body. The sleek fur was a dark beige color, with some variation around the muzzle. She described it as a large, agile, powerful wild animal that her husband feared could have hurt her, its weight being not much less than hers.

Another such story materialized in my neighbourhood the same year, and from all appearances, it was quite likely the same type of creature that I just described. It was not seen in Lake Manitoba, however, but in Lake St. Martin, which receives all the waters of Lakes Winnipegosis and Manitoba via the Fairford River.

It took place sometime in the 1950s before a highway was built to Anama Bay on Lake Winnipeg, when the Dauphin River was the only connection between the three Native communities on the shores of Lake St. Martin and the off-shoot community of Dauphin River. Oliver told me that he was a teenager when he drove a boat from Dauphin River to Lake St. Martin with his parents and siblings as passengers. His younger brother, who was around four years of age, was dangling his arm over the side of the boat, playing with the water as they traveled along. Suddenly a creature lunged out of the water towards the hand, narrowly missing it. It also had a dog-like appearance, and the color again was dark beige. It surfaced long enough that everyone in the family saw it, and their mother, who had been sitting near the young boy, reached over quickly and pulled him away from the danger.

I asked Oliver if he remembered any other references during his lifetime to a similar creature, and there were none he could think of. He said that his parents told others back home of the incident, just as if it had been a one-of-a-kind experience.

Such rare accounts these were, that to hear of two sightings seemingly of the same creature in size and color, virtually in my back yard, was extremely gratifying.

In Osik Lake, to the north of Nelson House, a trapper years ago, before the flooding, reported seeing dogs on a sandbar in the early mornings. He decided to try trapping them with double spring traps, but they would invariably be wrecked the next day when he checked them. He compared them to "wiener dogs" with big ears and short legs. In this same lake, one such animal was apparently caught in a fisherman's net, but it escaped. It too had been larger than an ordinary dog, and a howl was apparently heard from it.

A dog-like animal seen by Mrs. Kaartinen beside Lake Manitoba, and described to Jarmo Sinisalo.

It seems to be common knowledge in certain northern communities that these "dogs" play on beaches at night, and then go back into the water.

One veteran trapper had a simple explanation when I questioned him about how such animals could survive under water and under ice, with no noticeable regular surfacing for air. He compared their watery environment to what had surrounded them before they were born, and simply saw it as an adaptation to similar circumstances. His son made the same observation in regard to the underwater moose that is also believed to live under water and under the ice.

WHALES

IT RISES OUT OF THE WATER—usually on a calm, sunny day—and looks like a whale. So, until someone proves that it is something else, I'll refer to it as such.

A pilot called it that too. When he landed at Split Lake in north-central Manitoba, he said, "I didn't know you had whales in your lake."

Many witnesses, seeing it from ground or water level, describe it as an overturned boat or canoe. And the size? Anywhere from boat to bus to house!

When folks tell of their sighting, they usually add that the lake where they saw it has deep spots. Often, *very* deep spots for freshwater lakes.

Questions arise immediately. Aren't whales salt-water animals? And where are they in winter when the ice is very thick on all the northern lakes? Do they not need to come up for air?

These and other questions may have to remain unanswered for a while until more intense research reveals the truth, but, whether the stories are credible or not, there are an abundance of them. Some are dramatic, while others simply indicate the specter of an overturned boat in the distance, seen for a few moments, then submerging slowly. When it is not referred to as a boat or canoe that is upside down, it is called a reef, or big rock, sticking out of the water, out of place. And the shock is accentuated by the fact that the water is known to be deep there, with no known reefs in the area. But the reef inevitably goes down before the eyes of the beholder.

Boaters have been known to hit them, only to have their crafts beach so badly that they needed to push them back into the water. But, their efforts weren't required, since the "reef" submerged forthwith.

Another account maintains that a woman proceeded to use a "reef" as a bathroom, but the story did not elaborate on the consequences. It didn't need to.

The most common scenario involving these creatures is where a boat is speeding along in safe waters when a "reef" is spotted up ahead. The boat slows down, or begins to circle the object, only to have it submerge before their incredulous eyes.

God's Lake has deep areas, and it also has many whale stories. In fact, it almost seems that the creatures in this lake have a penchant for creating terror, making me believe that these are an especially vicious, malevolent bunch.

Witnesses describe the creatures as rising out of the water to a height anywhere from several feet to the height of a bungalow, and then they quickly submerge, sometimes repeating the process a number of times in quick succession, creating large waves as a result. One man witnessed a whale surging up and forward in an undulating fashion five times, causing tremendous waves.

The size of the creatures and the size of the waves combine to strike fear into the hearts of everyone familiar with them. The folks at God's River maintain that, in the days when canoes were used for transportation, the people would follow the shore rather than strike out across the open lake, even if it took several days longer to reach their destination. The story is told of ten canoes that set out across that open lake, unaware of its history, and only five made it safely. Half of them did not survive the turbulent waters caused by these creatures. Even with the larger boats of today, travelers are still afraid of being swamped.

Elders here mention a "whale" that washed ashore long ago, a caribou with horns in its belly! A vague reference was also made to a man once swallowed by one, but that may just have been a recollection of the Jonah story. But, I did hear similar tales in Alaska where a man and a horned animal had been swallowed.

One fishing guide from God's Lake Narrows stated that Americans saw and feared the same things as the locals: getting splashed by something big beside the boat, seeing a reef submerge in deep water, or watching loons pop to the surface all around the boat on a calm day just before seeing a massive creature rise above the surface.

An overturned 18-foot boat is how a lady from Nelson House described what she saw. Her house looked out on Footprint Lake where, one calm day, waves suddenly appeared as a large object rose three feet out of the water and remained stationary in that position for an hour or more. Then, as suddenly as it had appeared, it submerged, creating waves again. She remembered its shiny appearance, but no head was ever visible.

In Reindeer Lake, a fisherman had turned back after seeing an island outlined in front of him one foggy morning, in a place where no island existed. This lake is well known for its creature sightings, and it has some very deep spots in it. One part of the lake where the depth is many hundreds of feet it is named "Deep Bay." An outfitter, using a depth finder, had on one occasion seen an image that was about 14 feet long at a depth of 200 feet.

One lady from Brochet, which is also on Reindeer Lake, said that once, while using a depth finder, it had beeped and shown a depth of only two feet for a while, "as if something was checking us out," before the unit returned to the normal 400 foot depth. Imagine if the "something" had decided to put on one of its acts just then!

Another woman from Brochet told of seeing a big hump rise three feet out of the water on a calm day, moving forward as it repeated the up-and-down movement four times, sending out two foot waves like on a windy day.

At Kinoosao, a little settlement just inside the Saskatchewan border but also on the shore of Reindeer Lake, the big humps of this creature are a common sight. One man had been driving his boat on the lake, and had turned sharply to see where bubbles were coming from. Just below the surface he saw a body about the size of his boat sliding down into the water backwards. Its head was "over three feet wide with fins on it like a dinosaur." The eyes were about three inches in diameter. He remembers its color being black and grey and white and "so scary that I looked away and took off."

A girl from this community remembered seeing more than one of these large creatures rise and submerge simultaneously, and, on one occasion, a pillar of water had shot up into the air.

This phenomenon has been described to me on a number of occasions in a variety of communities, seen occasionally after a big creature submerged. In some cases, it was described like water coming out of a hydrant, blowing as high as a power pole. One girl, walking along a lonely stretch of shore, got a terrible fright when water suddenly blew high up into the air beside her.

A South Indian Lake woman tells of her terrifying experience with a big water creature. She was in camp beside a lake while her husband was out hunting when suddenly something huge resembling an overturned canoe rose out of the water. As it went back down, it splashed water all over the tent, which had a baby inside. The mother was so afraid that the creature might come ashore and hurt them that she banged furiously on a kettle to scare it away. She did not see it again, but the instant her husband returned, she clambered into the boat and insisted on being taken home.

Osik Lake has something fearsome living in it—a creature or creatures that may or may not be whales. Regardless what they are, they have the respect of all the folks living nearby in Nelson House. Some seasoned hunters/trappers/fishermen refuse to go out on this lake, especially after dark—and you can't blame them.

If you sit in your boat and can see its back on either side below you, then you know it's big. If you lose a gang of four nets, you know there's something powerful in the water. When its head is described as over two feet wide, you know it is to be feared. Then you hear of the family that tried to leave their dogs behind on an island in Osik Lake, but couldn't shake them with their three-h.p. outboard. As they turned around to go back, they saw five of the dogs being plucked from under the water,

STRANGE CREATURES SELDOM SEEN

"Overturned boat" seen in a number of freshwater lakes in northern Manitoba.

one by one. No wonder people believe there is some kind of monster in that lake. And to top it off, someone's boat was chased by a creature with a head two-and-a-half-feet wide with big, bulging eyes set on either side.

One fisherman claims that the creature owes him a total of ten nets, so he refuses to fish in the deep part of Osik Lake any more. Sonar has shown on more than one occasion that there is something monstrous there.

In Lake Winnipegosis, sightings of something big were quite common a century ago. A Mr. Oscar Fredrickson of Winnipegosis, who collected stories of sightings, wrote the following accounts which he entitled "Report on the Lake Winnipegosis Monster":

> In 1903, I lived with my parents on Red Deer Point, Lake Winnipegosis. Our house was situated about two hundred yards from the shore. About a mile south of our place lived a man by the name of Ferdinand Stark.
>
> One day Stark was down by the lake shore when he saw what he thought was a huge creature in the lake. It was moving northward along the shore, a short distance out.
>
> Stark wanted someone else to see the strange animal, and as we were his nearest neighbours, he came running along the lake to our place. All the while, he could see the creature moving in the same direction as he was, only going a little slower than he was running.
>
> Stark arrived at our home very much excited and breathing heavily, and asked my dad and mother to hurry down to the lake to see a strange animal in the water. By the time my parents got down to the lake shore, there was nothing to be seen. Whatever Stark had seen had disappeared. Looking somewhat bewildered, but still visibly excited, Stark began to describe what he had seen.
>
> All he could see of the creature was its big back sticking out of the water, and it was very dark or black in color. A number of gulls followed it and kept flying down to it as if they were picking at it. My parents did see quite a number of gulls still flying around.
>
> In 1935, Mr. Cecil Rogers of Mafeking and I made a trip to Grand Rapids on Lake Winnipeg. While there, I called on Mr. Valentine McKay, a resident of Grand Rapids for many years.
>
> As we were talking, the conversation drifted to strange animals that had existed at one time. To my surprise, McKay said he had seen some such animal in Lake Winnipegosis.

Following is Mr. Valentine McKay's story in his own words.

"In September, 1909, I was traveling alone in a canoe from Shoal River on Lake Winnipegosis to Grand Rapids on Lake Winnipeg. At the time I saw this animal I was standing on the lake shore. I had stopped at Graves' Point to make tea. I was at the edge of the bush getting willows for a campfire, when I heard a rumbling sound like distant thunder. As I looked out on the glassy surface of the calm water, I saw a huge creature propelling itself on the surface of the water about four hundred yards out from shore. A large part must have been submerged, judging by the great disturbance of the water around it. The creature's dark skin glistened in the autumn sun, and I estimated it was moving at the rate of two to three miles an hour. As I watched it, a member of the body shot up about four feet, vertically, out of the water. This portion seemed to have something to do with the creature's method of locomotion. The course it was taking was toward Sugar Island or Steep Rock. I watched it till it went out of sight. The number of gulls, hovering around this creature, followed it as far as I could see."

Mr. McKay said he had described this creature to a geologist by the name of Craig who said it was quite possible that it was a remaining specimen of a prehistoric animal that was once plentiful.

A great many people will think these two men just made up a story about seeing some strange animal or creature in Lake Winnipegosis. But, it is hardly probable that both men would think up a yarn about the gulls. Stark and McKay had never met, as far as I could find out. Stark moved from Winnipegosis about 1904 or 1905, and where he is now, if still living, I have not been able to find out.

In 1934, Captain Sandy Vance lived on Graves' Point. One day he was over at Sitting Island, which is on the northeast side of the point. He saw what he thought was a huge animal a short distance off shore. Vance said it was the biggest living creature he had ever seen in water. Mr. Vance had been captain on freight tugs for many years on Lake Winnipeg and Lake Winnipegosis. He said he had often seen moose and deer in water, so there is no reason not to believe that he saw some strange living creature. Captain Vance died some years ago. He was well known here.

There are quite a few Natives who have seen some strange thing. They call it "the big snake." Those of Shoal River claim that a monstrous animal was seen often off Sugar Island and Steep Rock in Dawson Bay during the latter part of the 19th century.

In September, 1955, something strange was seen in the harbor at Channel Island by Mr. Charlie Burrell and some American tourists whom Burrell had taken there to hook fish. When they first noticed this thing, they were fishing and the motor was not running. Burrell started the motor, wanting to get closer for a better look at this strange creature, but when the motor started, this thing moved ahead in a weaving motion and submerged. When they got to where this creature went down there was a dead mullet floating. It looked as if the creature went down the water after trying to feed. Mr. Burrell described this creature as having a row of disk-like fins along its back that stuck straight up. It had a weaving motion when going ahead.

In the year 1914, I was still living with my parents on the old homestead on Red Deer Point, Lake Winnipegosis. One spring day, I took my gun and went to the mouth of Fuller's Bay looking for ducks. I stopped in the rushes on the lake shore on the east point of Fuller's Bay facing west. It was dead calm, not a ripple showing on the water. The lake was open along the shore, and Fuller's Bay had no ice, except just in front of me, about 25 yards out, there were two cakes of ice. One was about 30 by 40 feet, the other piece was smaller, and was a few feet north of the big cake of ice. All at once, while standing there, I heard some noise in the ice. I looked up and saw that something was going under the big cake of ice. It was pushing the slivers of ice up so high, that some fell on top. This was on the inside edge of the ice. As this moved out, the icicles didn't come up so high. They were just sticking up out of the ice and staying that way. The icicles furthest out were only out two or three inches. At the inside edge of the ice there was only about three feet of water. Under the outside edge of this was deeper water, and whatever went under the ice didn't push on the ice where the water was deeper. There was only a narrow row of icicles pushed up. Something with a big fin or a disk on its back must have gone under the ice in order to push up the slivers of ice in the manner this did. The smaller cake of ice started turning around by itself. Whatever went under the big cake of ice disturbed the water enough to affect the smaller cake which was a few feet to the north.

I was just standing there wondering what could be in the water big enough to do what I had just seen, when a Blue Bill Duck came and sat just at the edge of the ice. I shot the bird and waded out for it. The water was to my hips. Whatever went under the ice was so big that it didn't have space between the ice and the bottom, therefore when it forced its way out to deeper water, it pushed up the icicles where the water was shallow.

The second time I saw something out of the ordinary was late in May, 1937. We were traveling down the lake in my gas boat. With me was Pay Nipinak of Camperville, his sister, four children, and my son Bui.

We intended to stop overnight at Moose Island harbor. It was late afternoon, about sunset, and the wind was very strong from the west, blowing off Dog Point. Pat was at the tiller, and I was standing up, watching ahead. Before we made the turn going into Moose Island, I could see an object about half a mile away, which I thought was a canoe heading for the same harbor. This thing was coming from the south, from the Papooses. As we turned at the turn buoy, I figured the canoe would get to the harbor about the same time as we would. I still thought it was a canoe.

It was cutting across in front of us, about a quarter of a mile ahead of us. Then I took note there was no man to be seen at either end. Still thinking this was a canoe, I figured the man must be lying down with his hand on the outboard motor. Then, when it was about 200 yards away, with the speed we were going, and the speed it was moving, I thought we would collide. By this time I could see it wasn't a canoe.

Still not thinking it could be anything but a boat, I said to Pat, "Turn to your left or you will hit this object." Pat got up and looked and turned the boat so we passed within about 100 feet. By this time, the thing was not moving, and it was down so low that the waved lake washed over it. It looked to me to be about five or six feet wide. As we were passing this, it appeared that the sides were moving in and out or up and down, loosely. I couldn't see any sign of a head. Both ends seemed to be a lot deeper in the water than the middle.

As we turned in toward the harbor, we could still see this object. It hadn't moved. There was a strong wind all night blowing on to the west side of Birch Island. Pat took a walk in the morning to see if this object had drifted shoreward, which it would have done if it had been a log or a boat, or anything that was drifting. I didn't bother going, as I felt sure this was a living animal. Pat didn't find anything.

During the summer of 1931, I had J. Saunders and Stoney Johnson working for me. We set nets between Little Channel Island and Big Channel Island on Lake Winnipegosis. The nets were new linen, and had never been used before. When we set the nets, the water was perfectly calm, and when we lifted them, we found a hole over 12 feet long in the third one. The question still remains as to what went through the nets seeing that there were no boats near this area.

In September, 1955, in this same area between the two islands, Charlie Burrell and two American anglers also saw a strange creature on the lake surface. It appeared to have a row of disks on its back, humps, and swam in a weaving motion.

One of the many stories told by the Indians of Shoal River tells of an Indian boy by the name of McLeod, who lived at the Red Deer River. He and his wife paddled their canoe out to a stray object near Steep Rock. As they came closer, they noticed that it was monstrous, and had a head that looked something like that of a horse.

Another strange story told by Beardie of Shoal River, tells of a camping trip on Sugar Island. He and his wife saw a huge animal go over the sand bar on toward the south end of the island.

During the winter of 1936, I got a vertebra in a gill net. The vertebra was perfectly round, five inches long, and three and a half inches across the ends, and I could put two fingers into the hole where the spine was. That winter my fish camp burned to the ground, and the bone went along in the fire. I had it in the window of our bedroom, intending to send it to Winnipeg in the spring. I looked all through the ashes after the fire, but could not find any part of the bone.

There was more than one bone in the net, but the rest broke away in the basin hole, just as I was getting the bones up on the ice. The vertebra was white, even whiter than the snow.

I described the vertebra to Dr. E. A. McLeaod of Winnipeg and he seemed quite interested. He said that if it was possible, he would send some skin divers to go out onto the lake with me. But, I have never again heard from Mr. McLeaod. If more bones were found, it might be possible to prove that some large fish or reptile-like creature did exist in Lake Winnipegosis.

The year I had the vertebra, I made the trip to Winnipeg to see if I couldn't find someone interested in going up the lake to try to find some more bones. I thought if more bones were found it would be possible to prove that some strange creature does still exist in Lake Winnipegosis and Lake Manitoba.

I went to the Winnipeg Museum. There a geologist listened to my story, but wasn't impressed. He said, "No doubt, you people see something, but it can not be a living creature as nothing like that exists in our lakes."

I said, "What do you think of the two men's telling about the gulls following whatever they saw?"

He said, "I was a little surprised at that story because gulls follow deep sea creatures that come to the top, even the whale. Creatures that move very little and stay under water most of the time gather bugs, blood suckers, and other parasites that cling to the hide. Some float loose and these the gulls pick up as well as picking some right off the hide of the creatures.

This bit of information proved to me that whatever Stark and McKay saw on the surface was some strange fish or creature that does not come to the surface very often.

It may not be in our time, but the time will come when it will be proven that something larger, much larger than any freshwater fish could and did exist in the 20th century.

This is the end of the accounts that Mr. Oscar Fredrickson wrote about the various sightings in Lake Winnipegosis he and others had made. Fortunately for us, he kept a careful record, believing that sometime in the future we would find out more about them. Thanks to him we have a good beginning, and my thanks extend to his son Bui as well, who generously shared copies of the stories not just once, but again after our house fire.

Not all the creatures mentioned by Mr. Fredrickson appear cetaceous, or whale-like. The one described by Mr. Burrell as having a row of disk-like fins along its back and traveling with a weaving motion sounds more serpentine, and since it was also described by someone living beside Lake Manitoba, it will be treated in the chapter on the Manipogos.

Most descriptions of the "whales" that I am aware of mention a dark color, with the odd one being beige, or grey, or even brown with spots on it. Charcoal or grayish-black seem to be the norm. "Short black hair" was one comment, and a few saw the tail break water, displaying a forked tailfin. Some, hearing bubbles rise from the depths, said it sounded like rain.

BIG FROGS

"How big?" you ask. I must confess that I am too embarrassed to come right out and say how big some of our northern frogs are purported to be. So, I'll just work into it gradually, starting from the bottom up.

Take the short lady who held up a frog by two legs. The other two touched the ground. Even if she were diminutive, that's still a lot of frog, isn't it?

Most people's initial reaction is about the size of the legs, would you believe. For food, of course. I suppose it's a more appropriate comment than the one they'd really like to make, but are too polite to say.

I agree that something this size is hard to swallow, but bear with me and note that this category of unusual creatures rates near the top as far as number of reports are concerned. And, knowledge of them is widespread. While certain other creatures appear to be more regional, this one is recognized throughout the north.

Where do they live? I hear the word "swamps" used regularly. And numerous accounts refer to their subterranean habitat—soft, swampy ground. One such story comes out of Nelson House where they were installing water and sewer lines for a new residential development. Apparently a backhoe dug up a huge frog that was seen by a number of people after being mangled by the equipment. Others were known to live under buildings, were seen hopping across roads, and swimming in the waters. The following story accompanies the green frog with black spots picture:

Claurie and her brother were out picking wild strawberries one day, and as she was bent over, she felt something brush against her hair as it flew over her head. Thinking that her brother had thrown something at her, she asked him why he did it. He denied doing any such thing, so as they looked around, there, a big hop away, sat this beaver-sized frog. It had a smooth "plastic-like" appearance.

Others describe them as being brown or golden in color, so a variation seems to exist. Those who have observed them hopping claim that they are able to jump about 15 feet in one leap.

STRANGE CREATURES SELDOM SEEN

The frog Claurie drew—which jumped over her head!

A man from Nelson House related his boyhood experience of traveling with the family one summer on the way to their trapping grounds. As he was wandering around, exploring as boys do, he encountered a big frog on one of the portages. He poked it with a stick until the frog grabbed it with its mouth and snapped it.

He also mentioned that a certain lake, known to have big frogs in it, has a reputation of being hard on boats. Apparently the frogs are believed to be the creatures that chew on the bottom of boats left in the water overnight.

The folks around South Indian Lake claim that there were many big frogs in their lake before the flooding took place. After that, they seemed to disappear, and many dead ones were seen floating in the water.

One old trapper from Nelson House reported seeing a big frog eating berries, pulling the branches down with its front legs. It was green with black spots and a white throat. He said its tongue extended almost a foot, and "it eats animals too." Some, he said, could be dangerous, and would even attack.

Some frog-related information out of Sioux Valley that I think is significant comes from an elder who made this comment, perhaps from personal experience. He had said that when certain frogs and snakes meet eye to eye, the frog gives off sparks "like a lighter" that can kill the snake. While this observation is neither relevant to the topic of giant frogs, nor totally unfamiliar, it is something of interest considering that we are talking about something Canadian.

Another sighting describes a frog reaching up and eating birch bark. However, there is something new here that I must tell you, and I hesitate to do so. . . .

You may have been inclined to believe what I have reported thus far, but I fear that my credibility and integrity may be seriously questioned hereafter. Understand, if you will, that these are just stories, without any material evidence to substantiate them at this point

Although the first such story I heard seemed like an outright myth, after hearing similar ones over a period of several years, from a variety of hunters in different communities, well. . . . skepticism tends to shrink a little more with each new account—a process which has repeated itself with each new creature that was brought to my attention during the years of story collecting.

That frog we were talking about, eating birch bark, of all things? It wasn't as big as a beaver. It was as big as a bear! And I wasn't just trying to abbreviate the word beaver, either. When you hear that a hunter leaned his gun up against a big rock, and then noticed the "rock" turn its head towards him, why, it tends to give real meaning to the expression "a hair-raising experience." Many are the persons I talked to who described the sensation of hair on the neck standing up, and it didn't happen only in connection with big frogs.

The sensation was probably similar for the man who had sat on a "rock" only to have it jump into the water.

Near Pukatawagan, two firefighters thought they were just looking at two small boulders side by side on a rocky hill one July, until they came within 25 feet of them and saw their heads move. One of the men, Gabe, described them as being toad-like, grayish in color with darker spots, and about three feet every which way.

A similar-sized frog was seen with a ballooning throat, making calls that echoed a great distance.

The story goes that some firefighters were using so many pumps to pipe water out of a small lake that the water level dropped noticeably. Objects that looked like boulders moved towards deeper water, betraying their identity. The men apparently took extra precautions that night to make sure no big frogs entered their sleeping quarters, by using stones to hold down the perimeter of the tent.

Brian Wood's sketch of the big frog he saw.

A gentleman from Nelson House described what he remembered seeing with his mother when he was a boy. The huge head of an olive green frog had been visible in the water, its eyes six inches in diameter. The head he compared in size to the end of a 45-gallon fuel drum.

Then there were the hunters who heard something big coming out of the woods to the water's edge, expecting it to be a moose. Instead, it turned out to be a bear-sized frog.

Perhaps that is the creature that travels under the surface of bogs and sometimes lifts it up as it moves.

Now that I've gone this far, I might as well go the whole hog. What more do I have to lose? Don't experts say that reptiles never stop growing? Well then, how about amphibians? What if some of them kept growing ... and growing ... and ...? How big could they get? Could a frog really get as big as a moose ... or a shack ... or even bigger?

If you can afford to hire a helicopter on a calm, sunny day when you can see to the bottom of lakes in the north, fly over some of them and see if there is any truth to the stories of frogs as big as ...

DRY RATS

THESE ARE SO CALLED because they live in dry swamps. So what's the big deal with a band of rebel muskrats that decided to break with tradition and leave the wet swamps behind? Perhaps their leader had a strong aversion to water and led other like-minded rats to higher ground.

Somehow I think it goes deeper than that.

The name may sound innocuous enough, but to some folks it is the stuff of nightmares—the worst of nightmares.

A little muskrat, you say? What's going on here?

Well, here are a few notes from those who lived to tell the story. No, they didn't tell me. I just got it second hand you might say—except for in one instance, which I'll tell you about later.

Out of Berens River comes the story of a man who happened to cross a swamp where these muskrats lived. Several started to chase him, and as he stole the odd glimpse while he was desperately trying to get away from them, he noticed that they were not running as much as they were jumping. He also saw that there were caribou bones all over! But, he survived to tell the story.

A similar experience comes from Poplar River, a community in the same region as Berens River, on the eastern shore of Lake Winnipeg.

Four men were crossing a "dry swamp" when one rat attacked. It jumped high, towards the face. Of course, it was outnumbered and over-powered, but can you imagine what might have happened if it had recruited its whole village first?

Are they rabid, or part wolverine?

Elders advise anyone encountering them in late spring to take off a snowshoe quickly and use it as a weapon. Probably as a shield *and* a weapon.

One lady I spoke with told me that a creature resembling a muskrat had once jumped up to her shoulder, so she had killed it. Simple as that. I don't think she had ever heard of "dry rats" before.

A man hinted that one such rat was capable of wiping out a whole camp. But, he insinuated that a curse was behind it.

The consensus seems to be that this unconventional, aggressive animal is a vicious little beast that quickly attacks any intruder, going for the jugular and not letting go.

Those caribou bones might just be the proof.

BIG MOOSE, SMALL MOOSE: GIGANTISM AND DWARFISM

As with some of the other unusual creatures I was told about, I didn't pay much attention to the first super-moose account, but when I heard it mentioned again, the stories began to take on more significance. Furthermore, I found information on the large prehistoric moose that were known to live in North America long ago, so I wondered if there might be a connection.

Remains of two species found here in North America are the broad-fronted moose and the slightly smaller stag moose. The broad-fronted moose was apparently larger than the Alaskan moose of today, so the giant moose stories I hear may just indicate that another seldom-seen creature, like the giant beaver, is not extinct after all. Let me share the few but interesting accounts I have heard to date.

At Split Lake a man told me that his father had once seen a large moose, but it was not until he got close to it that he spotted two ordinary bull moose nearby that gave him a size-comparison—and a shock. All had carried racks, but the two smaller ones "looked like dogs" in comparison.

A similar, but first-hand account from Pukatawagan is somewhat more dramatic. Angus and Pelangie Linklater were hunting by canoe east of their community, about halfway between it and Nelson House. Beside a little river they spotted a big moose bearing a large rack, so they paddled closer in order to shoot it. As in the other story, it was then that they noticed another, smaller moose which also had antlers, harassing it. It would come up to the bigger moose from in the bushes behind it and begin to charge, but at the last minute the larger one would wheel around, causing the smaller one to retreat. Angus said that seeing the two together made the ordinary one look like a calf. He was inclined to move closer for a better shot at the huge animal, but his wife was so terrified at the sight that he let it be.

About a year later, Angus encountered a huge grayish-colored moose again, near enough to the original site for it to be the same creature. Again, a smaller moose was near it, so he brought it down with a rifle typically used on moose. But when the larger beast stood its ground, facing the occupants of the fishing boat just a short distance away, Angus grabbed a bigger rifle and knocked it over with one shot to the forehead.

As he described its massive size to me, he reached up towards the ceiling of his kitchen, explaining that he was indicating the height of its back only, not the top of its huge head.

On my recent trip to Pukatawagan in the winter of 2011, Angus recounted his story again. A man who had been with him on that eventful trip dropped in and repeated what Angus had told me all along—that it had taken three men to carry one hind quarter! Usually, they said, one man could handle it. And as for the head with antlers—it was so difficult for four men to move that they just left it behind.

When I mentioned this story to my friend, Doug Reader, who had grown up at The Pas, in northern Manitoba, he recalled his Dad's report after a moose hunt one winter in the late 1950s. In a large willow swamp north of The Pas, they had counted nine bull moose together, but one had stood "head and shoulders above the others." All were disappointed that his Dad's .30-30 hadn't been able to bring it down.

I found the first miniature moose story equally unbelievable. It originated at Oxford House, a reservation in Manitoba's northeast, so remote that it took me many hours on the winter roads to reach it.

A gentleman from there told me that the last one was seen in the 1950s. How big? Not. The size of a deer. What distinguishes them from a moose calf? Antlers and tough meat.

As I was discussing creatures with an acquaintance from my neighbourhood, he remembered the story of someone who had gone out trapping but didn't want to kill a large animal. He had encountered this small moose that fit the bill, but he discovered that its meat was like that of a tough old animal, so he concluded that in spite of its diminutive size, it must be an old one.

From the same area a similar story has been passed down, and this one involves the man who habitually fed the Manipogo in Lake Manitoba—Andrew Sumner. He and his partner had been out hunting in the north Interlake, and upon returning to their camp at the end of the day, they saw fresh moose tracks. When they shot the small, deer-sized moose, they were disappointed that its tough meat was hard to chew.

The setting for the last two stories above are essentially in southern Manitoba, so it will be interesting to see if any similar accounts surface anywhere else on the continent.

Perhaps these may just be stunted moose, or mutants, in the same category as the giant moose. The very few reports of both huge and tiny moose would seem to confirm that possibility, whereas among other giant creatures where hundreds of sightings are known, they may in fact be species altogether different from their smaller counterparts.

Gigantism is not a foreign concept to most Native people. One lady from Easterville said that her grandfather had told her mother that each species of animal had a giant counterpart. She had no difficulty believing her grandfather's theory, for she and her daughters were the ones who saw a giant beaver cross the road, and it was her husband who saw the giant golden eagles fly off with a deer.

Many Native people are quite familiar with giant snakes, big fish, huge frogs, and a host of yet unidentified monsters in various lakes. Even the Sasquatch is a giant in its own right, one that has been witnessed probably more frequently than any other little-known creature.

Dwarfism is perhaps not as common, but it does deserve some mention, considering the reports of the miniature moose.

MERBEINGS

IF THE QUESTION of whether or not merfolk exist depended on the number of entries in my notebook, then I would point out that they rank among the top six creatures. So, if I have perhaps convinced you that big beavers, big snakes, Sasquatch, Manipogos and whales exist in Manitoba, then you might believe in these gentle beings as well. After all, have you not seen pictures of them and read stories about them throughout your life?

I felt the same skepticism towards them as when the platypus was first mentioned to me, and, although I don't expect to see one with as much hope as I do the platypus, the sheer numbers and similarities of merfolk sightings is overwhelming.

I see again the radiant, excited face of the young lady from Pukatawagan who, unsolicited, came forward with her experience when we found ourselves together in the house of a man I had gone to see about other creatures.

When Lorna was 14, she and four other girls had enjoyed diving into the lake to see who could swim the farthest underwater. On this occasion, she had seen a "mermaid" in the clear water, with long golden-yellow hair. She feels that they both startled each other about the same time, but her eyes were riveted to the beautiful hair. Lorna said that the face reminded her of a human, but she recalls the fish-like posterior as terminating in a pair of flukes that were close to three feet from tip to tip. She said it was the last time she went swimming away from the shore.

Not many reports are quite as concise and dramatic as this one, but the similarities are convincing. And the number of entries from Pukatawagan in this category makes it by far the richest community for sightings. Some reports simply consist of so-and-so seeing a mermaid, but others, as you saw, are more specific. One boy stated that when he was swimming, he had seen what looked like a human head with long hair sticking out of the water.

However, another merbeing story from Puk has me perplexed and a little disappointed, simply because it doesn't match the majority of reports, but since it came from a mature, articulate woman, I have no reason to doubt her story.

Joanne was 10 or 11 when she too was swimming in the lake, and saw a form from very close that was "naked, and exactly like a human except for its webbed feet and hands." She saw buttocks, and the whole body had smooth skin "like rubber gloves—a pale blue with whispy green pieces attached." Do you see why I'm disappointed? But if I ever hear another report like this one, I'll get excited again.

So far, every other mention of the creature has included a fish-like bottom, not legs. And the color has generally been white, with the hair yellow or red. The majority of witnesses have observed female characteristics, but not all.

A pastor's wife who grew up at Poplar River on the shore of Lake Winnipeg saw one when she was going for a long walk along the water's edge with two other girls. She told me that they had seen what looked like a white woman sitting on a rock in a distance, and wondered what she was doing there. The girls weren't able to follow the shore all the way, so they surprised the creature when they looked down on it from close-up a little while later. But before it dived into the water, they saw what looked exactly like the top half of a woman with long red hair and a fish-like bottom—with her hands on her lap.

Although merfolk stories are few from where I live along the Lake Manitoba, Fairford River, Lake St. Martin, Dauphin River, and Lake Winnipeg water systems, there are some reports worth mentioning—like the large pinkish flukes seen above the water in a deep area of the marsh, belonging to something larger than the local fish. Another local resident told me that his father, when they were in that area of the marsh where merbeing sightings had been reported, had made the comment, "This is where your girlfriend lives." He hadn't known what his dad meant at the time.

This correlates with a report from a man I used to know at Fairford, regarding a fisherman who once caught a child-sized merbeing in a net, and it had long red hair. A lady from the same community is to have lost something from the end of her fishing line, but brought up long red hair on her hook. A report from Puk mentions something being tangled in a net under the ice, and it had long brown hair.

Someone's grandfather on the east side of Lake Winnipeg knew about a woman-like being that was caught in a net. The fisherman had apparently talked to it and then let it go. A report from Moose Lake alleges that a man from there caught a small "ugly" mermaid in a net.

Back at Fairford—a man who lived beside some rapids claimed he saw a mermaid early one morning, and it had been combing its hair. In the neighbouring community of Little Saskatchewan, some young folks "stole" their neighbour's boat and paddled it along the shore. Looking over the side they saw a six- to seven-foot-long object with its head behind a stone, its flukes gently moving. Its body was the size of a human, getting slender towards the tail. It was a teal-green color, "just like Ariel in the mermaid story. It was too big for a fish."

Eden Lake is situated between Leaf Rapids and Lynn Lake. It seems to have gained the reputation of harboring merfolk.

A unique sighting comes from Oxford House where merbeings are well known. A man told me that he witnessed two human-like creatures passing a fish between them above the water before they went back down. Another resident of this community described what he saw in the water when he was nine years old. It looked like a white woman with hands just like ours, but it had a tail like a big fish, and long brown hair. The sight had scared him, but when he told his dad, he hadn't been surprised at all.

One lady told me that mermaids were even used as a threat to control children's behavior. She said at Lake St. Martin people would tell children not to play in the water or along the shore "or a mermaid would get them."

Only three reports mention merbeings being seen combing their hair, and that was near Big Rock on Lake St. Martin, in the Fairford River, and at Oxford House.

As you may have noticed, the majority of merfolk are seen sitting on rocks that are surrounded by water, and often this is near rapids.

Several men associated these creatures with tangled nets and missing fish, while others, seeing ripples in the water, assumed that the singing or laughing they heard originated from something that lived in the water.

If it were only men who saw these predominantly female creatures, we might tend to dismiss their reports, but as you can see, a good number of ladies told me about their experiences.

I recall hearing more than one account of merbeings following men in boats, and the head of a creature that popped out of the water near some rapids almost every time a man I know went by, could very well be one and the same.

MISCELLANEOUS CREATURES

IN THIS CHAPTER I would like to mention the various sightings and stories that do not have the same body of evidence surrounding them that some of the more prominent animals do. It is not that I doubt the integrity of some witnesses, but if theirs happens to be the only report I have heard on a certain creature, then I tend to put it on the back burner until such a time as more corroborating evidence surfaces. You will find just such rarities in this chapter. But in addition to them, I will include the creatures that are not seen clearly enough, or were not out of the water enough, to be given any definitive description. And, there may be the odd "you've got to be joking" animal that has the potential of tainting the whole book!

To leave any of them out, however, would be to leave that many stones unturned. Who knows what future research might uncover that today may seem far-fetched or ridiculous. So, believe them or not, enjoy the "What on earth?" section of the book—if you haven't encountered it already!

HIPPOPOTAMUS?

THE WORD "HIPPO" has steadily gained prominence in my notebook as a creature whose appearance loosely matches what a number of eyewitnesses have described. The smooth broad back is seen in various northern lakes, but the most complete description comes from Lake Winnipegosis where a man's uncle had stated that the animal he saw belonged in the jungle. The huge snout, large nostrils, small ears, and heavy round footprints certainly suggest something similar to a hippopotamus.

A man from Split Lake who had seen a hippo on television said that what he saw in the water much resembled the back of it, and the expansive snout and small ears matched as well. Similarly, a man from Pukatawagan extended his arms to indicate the broad back of what he saw. In the summer of 2009, a lady and her 12-year-old son were fishing from shore near Pukatawagan when they noticed a pair of eyes the size of fists protruding from the water. Behind the eyes, just below the surface, was a smooth-skinned body about two feet in width.

Another man from Split Lake had once seen some strange footprints in about two feet of snow near the Nelson River, and they had reminded him of something heavy like elephant feet. The "frying pan" sized prints which were about four feet apart from side to side also clearly indicated an animal that dragged its feet so much that it scraped the snow right down to the ground.

There are some questions that definitely surround this shy aquatic creature. If it has feet as in the above story, then why aren't more tracks noticed on shore? Unless, perhaps, it, like the giant beaver, has little need to be on land. However, what it would do in winter under the ice is anyone's guess. Perhaps it doesn't require an ice-free area to come up for air, but like the amphibians (and the underwater moose!), it just beds down quietly somewhere on the bottom of the lake.

WILD PIGS/BOARS

I HEARD ONLY THREE REFERENCES to creatures that could fit this category, and all were seen in the eastern half of the province. One boatman had reported seeing two "wild boars" that he followed in Lake Winnipeg. One trapper had witnessed several "pigs" come out of the same lake, and the other sighting was in the Canadian Shield near the Manitoba-Ontario border. In this case, five men had seen a herd of black and white pigs in the wilderness, and they had run into the water when the men tried to catch some. One of the witnesses told me that they had never heard of them before.

Whereas the above stories involve cloven-hoofed wild boar-like creatures much like those known to live here and there on earth, the following stories are about something similar and yet uniquely different in several respects.

Dennis, from Pukatawagan, visited his remote trapping area by canoe one summer, and awakened an animal that had been sleeping on shore. As it ran, it made a vocalized sound continuously, much like domestic pigs do when they run. It was black in color, but its hair, or bristles, had not been thick like fur. The outstanding feature that Dennis noted was a seven or eight inch tusk that protruded from the end of its snout. Furthermore, when he checked its tracks, he was in for one more surprise, as it did not have split hooves. Instead, its footprint was round and flat, "like the heel of a man's boot."

Another interesting observation that Dennis made was that nearby was a creek that never froze. When I shared the information with Angus, my host, after I returned to his house, he showed no surprise, but matter-of-factly stated that something unusual could be expected in an unusual environment.

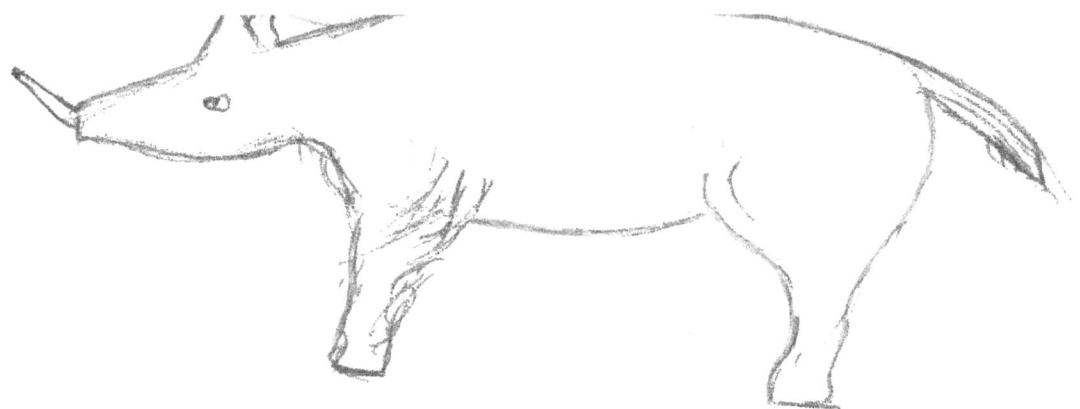

Pig-like creature seen and sketched by Dennis Dumas of Pukatawagan.

I didn't really expect to hear a similar account from anyone else, but within a short time of hearing the first one, I heard two more, simply because I was now aware of this creature, and could specifically ask other trappers if they had ever heard of it. When I returned to Nelson House and visited again with Joshua, who had already shared numerous experiences with me, I mentioned the "pig with a tusk" story out of Pukatawagan. He said that he had encountered the same animal in his trapping area, and, without knowing all the details of Dennis' sighting, he mentioned that the area where he saw it was swampy and did not freeze. That detail was the clincher for me, as I now knew that the two stories had all factors in common.

The third account was not first hand, but similar enough perhaps to be the same creature—and near home in the Ashern area, of all places.

MONKEY/KANGAROO/SQUIRREL?

A SURPRISING NUMBER OF REPORTS indicate a monkey-like figure that scoots across the road very quickly, leaving the witnesses rubbing their eyes in bewilderment. The word monkey is always used, but squirrel and kangaroo are sometimes included to indicate some aspects of its appearance. The sizes vary, but usually they are described as being considerably smaller than a human.

One recent sighting on Highway Six near Devil's Lake was made by two sisters from Jackhead. A monkey or ape-like animal ran fast across the highway about 200 feet in front of them. It looked to be about four feet high at the front, and its neckless

Creature seen and sketched by Michael Sumner of Fairford.

Boyhood recollection of a creature seen near Fairford by Curtis Woodhouse and his uncle.

back sloped down obliquely like that of a typical chimp or ape. This one appeared to weigh about as much as a human, and its fur was a light brown color.

In the summer of 2009 a man was traveling alone on the highway north of Pine Falls. He stopped to look at a creature he at first thought was a rare brown bear. However, he told his friend that it looked more like a kangaroo, and hopped away just as a kangaroo would. Unfortunately, the man passed away shortly after, so no additional information is available as far as I know.

MONKEY MEN?

A VARIATION OF THE ABOVE CREATURES are ones that are legendary in the southern parts of Alaska and Yukon in the regions where the two meet. As I drove through the area, they were mentioned by several individuals, and it was recommended that I go to see certain people on the Tetlin Indian Reservation near the Yukon border. I am very glad I took the long, winding road to the settlement, for it was there that I got some first-hand information about the monkey men.

But let me first share what I had heard earlier about these unusual creatures. It was believed to be before the Russians came to their coasts in the 1700s that a significant conflict developed between the Natives and the monkey men. The story as related to me went something like this:

A group of Natives traveled to a favorite fishing spot to stock up for the winter. When they did not return, a concerned individual went to look for them. He encountered a group of monkey-like creatures that were happily playing a game of sorts with a round ball. As he watched the activity from his hiding place, he was horrified to realize that the ball was in fact a human head that belonged to one of the people he sought. He hurried home and organized a war party.

Since the creatures lived underground, the warriors chased them into their holes, and then filled the openings with wood in order to burn or smoke them to death. This tactic was apparently effective, since there were no more known incidents involving these beings again.

As I said, this was the version that I heard by word of mouth. Later, after I purchased the book *Strange Stories of Alaska and the Yukon* by Ed Ferrell, I found an account about "The Tail Men" that was similar enough for the two stories to be one and the same. In fact, given that the information was passed on to a credible individual who recorded it almost a century ago, it is likely to be more accurate than what I heard. However, this may not be the end of the story. When I asked about unusual creature stories upon my arrival at Tetlin, I was referred to a young man who told me the following story:

The monkey-man of Alaska/Yukon, sketched by witness Lee Joe, Jr.

Around 1980, when he was a boy, Lee Joe, his mother, and a number of others were returning on foot one evening from berry picking. As they came over a ridge near a swamp, Lee, who was in the lead, got a close look at a creature that was digging into the vegetation. When it saw him, it made a piercing shriek, and with a few bounds disappeared into the bush. Joe made a quick sketch of it for me, and noted that its eyes had been a deep red color with black centers. Its shriek he compared to the sound of fingernails on a chalk board. Where it had been digging when he first saw it, there were two muskrat-house-sized piles of vegetation.

The next morning when I visited with his mother and her sisters as they worked with the spruce roots they had dug up the previous day for use in sewing their birch baskets, I showed the man's sketch to his mother. She agreed that it was fairly accurate, but that the tail might have been a little longer. She said the hind legs reminded her of a kangaroo's, and the creature, which was a little smaller than a human, was a light brown color. The sighting that she was party to apparently was an isolated incident, and I heard of no other such present-day stories.

An elderly lady in the community told me that her mother, who had been born in the 1870s, heard tell of the monkey men, and around 1900 had been shown nine holes in the ground where people had burned them out. At that time the holes had already begun to deteriorate, but the fact that they were still there indicates a fairly recent time frame for the conflict. It was when she first saw pictures of monkeys that she had been reminded of the stories.

I heard several individuals say that the creatures had looked like people, except that they had tails. If that is the case, then the one seen around 1980 would not qualify as being one of them, and yet its monkey-like appearance does match the references to monkeys that are made in the stories.

The lady who saw the holes also told her daughter that a man once tipped his canoe in a certain lake, and held on to it while others worked on building a raft to rescue him. He shouted that something was eating him from below, and he succumbed before he could be rescued. A short, snake-like animal was thought to be responsible. Perhaps it was the same scenario where a man went into a lake to retrieve his moose, and came out covered with "bloodsuckers." He too died as a result.

RECENTLY EXTINCT?

As I spoke with various people in my travels through Alaska and Yukon I got the impression that some creatures believed to be extinct today were very much a part of the oral traditions of the elders. If, in fact, some of them are extinct, it may not have happened all that long ago.

One man told me that some elders talked of a "hunter with knives for teeth" which he took to mean the saber-toothed tiger. Huge land animals thought to be the woolly mammoths are also mentioned. In fact, he felt that these creatures may still have been around about 500 years ago. However, from some stories I heard and read from that region, I have the distinct impression that the mammoths have not been gone even that long. A camel-like animal is also well-remembered, especially because of its strange aggressive habit of going for people's privates.

Among the stories told to me by Dawn Charlie in the Yukon is one from her father-in-law who claimed to have seen what his people refer to as "beaver eaters." Apparently they are bigger than grizzlies, have long claws, and use them to open up the tops of beaver houses. These slow-moving creatures, thought to be ground sloths, were mostly seen on a certain mountain, and the last sighting of one was about 70 years ago. Some believe that they may still be there. One man had said that a beaver house raided by one of these creatures looked like it had been dynamited. "Beaver eaters" is a name that has been given to the wolverine in some regions.

People from this particular area remember being told about the mammoths, including the specific place where the last one was killed "because it was a threat to humans."

This area has a story like the one from northern Manitoba where a hunter's second caribou disappeared from beside the lake before he could get to it. I met a couple living near the Alaska-Yukon border, and from them I learned of an incident that involved Roy's parents. Sometime in the 1920s they went hunting near a lake they called "Killer Lake." Leaving his wife in camp, Roy's father went out alone and shot two big bull moose at the edge of the lake. He went back to camp to get his wife to help him pull them out of the water, but when they got back, the moose were gone, and all they saw was swirling, churning, bloody water. (By the way, it was Roy and his wife Avis who told me they saw a giant beaver once when they were out hunting together. They went towards it with their canoe, hoping to shoot it, but it disappeared before they got close enough. There had been no house, and Roy was certain it lived in the bank.)

As in Manitoba, I found that there were a number of lakes with monster stories attached to them, some of which were of such proportions that I hesitate to elaborate. This is also the region where stories of Sasquatch are frequently linked to abductions, especially of women and girls (boys are generally released or returned!), and this is where they are believed to have some power over people's minds. Whistling in the dark is very much discouraged.

MISTER SPENCE

When there is not an obvious name to attach to an unusual animal, the name of the person most closely associated with it comes in handy. I met this gentleman at his home in Nelson House after a number of people suggested that he had a story I would find interesting. All called him Mister Spence for some reason, so I will too.

He and his partner were hunting some years ago in the area of Wuskwatim Lake where a power dam is being constructed today. They saw something that defies the imagination, and if it weren't for the fact that many folks heard about the encounter from both men, and recommended that I get their story, I would be loath to share it with you.

The only living witness is Mister Spence, and he described his experience to me on several occasions. He and his partner set up camp on shore near a small body of water. Something swam to shore near them, and stood up on its two long legs with the help of its short "arms." When it became aware of them, it put its long neck down and disappeared into the water. They saw its entire body long enough to make the following observations: it stood about ten feet tall and had a similar body size to that of a human, with similar-sized legs as well. Its front feet were quite short, however, and had hands something like ours. It had several large crest-like appendages on its back, and a substantial tail that was about six feet long and touched the ground. It had only a small head atop a long neck that was about as big around as a human's, and its color was grayish. The men found a hole nearby where they thought it lived, and the diameter was almost four feet.

Mister Spence told me that he had never heard of such a creature before, and I also thought for some years that this was an isolated report. However, as I read through my notes recently, I noticed a less-detailed description of a "humanlike" creature that may just be the same kind.

FURRY SNAKE?

This next story is definitely one of a kind. A man from Kinoosao, on the Manitoba-Saskatchewan border west of Lynn Lake, told me he had been working with a survey crew at Baker Lake, Northwest Territories. He was carrying a heavy drill on his shoulders and was just going to cross a chest-deep creek when he scared up a snake-like creature that slithered away on top of the water very fast. It had been about a foot in diameter and a dozen feet long. The really odd thing about it was that it was covered with brown fur "like a winter hat." He figured it went about 100 yards in several seconds. He was so scared of its return that he did not cross the creek, and therefore got into trouble over it.

This unusual biped was seen and sketched by Mister Spence.

STRANGE CREATURES SELDOM SEEN

GRANVILLE LAKE MONSTER/DRAGON?

ONE MAN FROM GRANVILLE LAKE told me of a large creature he saw as a boy. He used the word "dragon" to describe what he saw of its head and part of the body, and with his hands he indicated the length of its head at about four feet, and the width of it between two and three. He also mentioned that it had horns! When he had told his dad about it, he was not surprised since others had reported seeing it as well. I find that the word "dragon" has been used by several witnesses to describe the head of a creature they saw in various locations in the north.

One spot that is perhaps worth the mention is Pauingassi, a small Native community on Fishing Lake near the Ontario border. The first time I heard the word "dragon" mentioned, I didn't take it too seriously, as it was used by a young man who was slightly intoxicated. He did, however, make a remarkable sketch for me, which I just filed away, and showed to very few people. Then, some time later, I was speaking with a gentleman from Berens River who had been at Pauingassi as a firefighter, and he became animated when he saw the sketch. He had not seen the creature, but when he and other firefighters had camped beside the lake, they had been kept awake by a strange, loud, goose-like call. When they had asked the locals about it the next morning, they had been told that the sound came from a "big snake." The firefighters then asked to move their camp to a different location.

ASSEAN LAKE MONSTERS

GO SEE JOHN GEORGE BEARDY, I was told by people at Split Lake. This happens all the time. Neighbours, relatives, or friends hear what so-and-so saw, and I get dispatched to collect the goods. It's a perfect system. So I go and find his house with the door padlocked, all day . . . and the next. The following year his name comes up again, and, finding that he has a telephone, I finally catch up with him and enjoy a long conversation about some of the incredible things he has seen. I go see him face to face next trip I make to Thompson, going the extra hour and a half to Split Lake.

The first order of business, of course, was the big one he and his son Melvin saw in Assean Lake around 1995 when they were transporting lumber by boat for their cabin. The lake is just across the highway from Split Lake to the north. They thought at first it was a big stump sticking out of the water, but it had a head and big eyes, and color. A short time after it had gone down, it reappeared, and they got to see it again. John George was very specific about the way it went down both times. Whereas many creatures seem simply to submerge, this one went under in a diving manner with a splash.

They were within 100 yards of the monster, and estimated its diameter at an incredible two-and-a-half feet, with about six feet of the neck being above water.

Assean Lake monster seen by John George Beardy and his son Melvin.

The head was even larger than the neck, and the big black eyes were about eight inches in diameter. The throat area was a dark yellow, but the rest of the neck was like a green and brown camouflage mix, with grey on top of the head. After it went down the second time, they felt their boat heave as if it had swum just below it.

John George's grandfather and some others mentioned seeing a very long creature in the same lake, but who knows if all their sightings might not be one and the same. Assean Lake certainly has the reputation of harboring a variety of creatures. The long one reminds me of what a gentleman from the same community told me years ago that he had seen in a lake from the air; it sounded like something serpentine that was over 100 feet long. Sonar had shown perhaps the same thing in Osik Lake—something of a similar length.

SPLIT LAKE MONSTERS

IN NEARBY SPLIT LAKE, where the community is located, a variety of monsters have been sighted besides the ones that fit the category of whales. Some have heads comparable to those of moose, and some are even larger, belonging to creatures in the order of the Assean Lake ones.

One man told of seeing something swimming slowly, near dusk, and the awesome size and sight of it made him want to turn away. It displayed a long, vertical neck, but the side view of it showed a large elongated head and snout. It was also seen from about 100 yards, and its one-and-a-half foot diameter neck was out of the water about eight feet. Although its size approaches that of the one in Assean Lake, its head seems to be altogether different.

Firewood used to be needed in large quantities in the community, so men would float it home from the shores of the Nelson River by pushing two or three rafts of it in front of boats. On one occasion, two men were in their boat pushing two rafts ahead of them during the calm of night when they noticed, with the help of the northern lights, something large circling them. One of the men became so unnerved by its presence that he climbed out of the boat and up onto the raft of firewood. It swam near them for about an hour, but they couldn't tell what it was. Apparently it wasn't quite as big as a moose, but it appeared to be covered with hair. A similar experience that was related by George Kechikeesic's dad included the mention of the effect the creature's movements had on the wood. The 16-cord rafts that were tied to a boat rocked when the animal swam beside it.

George mentioned an experience that he and his dad had that sounds so common throughout the north. They were traveling by boat when they hit something, but nothing broke, and no marks were visible, so they felt it must have been a large animal rather than something hard.

Normally, larger tracks seen in the snow in this region would consist of moose and caribou, but rumors of some huge and unusual ones are heard from folks at Split Lake as well. In the same vein as the frying pan-sized footprints I mentioned earlier, there are the snowshoe-sized ones, but an interesting feature added to this account is the indication that this creature dragged a tail.

John George's dad told him of a trail he had once seen in the snow. It was "like a worm that had made a path two feet wide, and where it had gone close to a tree, the bark was rubbed off."

Another man's father passed down a story from an old man who had been beside a deep lake up north that was named "Scary Lake" in Cree. He had seen a big creature come out of the lake and go into the bush with the use of its big flippers and big long tail. He never went back to the bush after seeing it. When he saw Godzilla in a picture or movie, he remarked about the resemblance.

Perhaps this is the same monster that was known to cross over land from one lake to another, flattening a trail that was compared to one made by a bulldozer.

An occasional sighting in Split Lake involves an animal that has very long hair on its back, has a big snout, is eight to ten feet long, and is at least three feet wide. This may very well be a match for what was seen circling the rafts years ago.

Some folks from York Landing, which is another community on Split Lake, mentioned another phenomenon that appears occasionally in their waters. It is a

Footprint Lake and Split Lake sightings.

Footprint Lake and Osik Lake sightings.

whitish, triangular form that appears briefly and then submerges. It sticks out of the water about a foot and a half, and is almost twice that long. While it may be the vertical fin on a water creature, what men from Poplar River on Lake Winnipeg saw was probably something different. Their horizontal white shape reminded them of a flat kite that dove under their net, avoiding capture.

Perhaps a similar creature was disturbed near shore when a lady at Oxford House pushed off a boat with her paddle. She poked a black object that she thought was a rock, and it swam away with a flapping motion along its edges. She described it as two to three feet wide, almost two feet long, and three to five inches thick, with no tail. She asked me if we had stingrays in Canada!

LAKE WINNIPEG MONSTERS

GRAND RAPIDS, MANITOBA, is a side-by-side town and Native community where the Saskatchewan River empties into Lake Winnipeg almost 300 miles north of Winnipeg on Highway 6. Besides being one of Manitoba's Sasquatch hotspots, it also seems to harbor water creatures that are out of the ordinary.

Russell Scott told me his story on more than one occasion—after I had initially been sent to talk to him. He and a partner were out in Lake Winnipeg one summer lifting their nets when they saw something large in the waters of a bay about three-quarters of a mile away. Thinking it to be a moose, they decided to go and catch it, so as one man drove the boat, the other got rope ready. When they got within half a mile of it, it must have seen or heard them coming, for it went down so quickly that water splashed up quite high on both sides of it. It appeared to be black in color, and four to five feet wide, with about two feet of its body out of the water.

Other parts of Lake Winnipeg have had their share of sightings as well. One man from Bloodvein told me that what he and his partners saw from a large boat one night in the bright moonlight somewhat resembled the large head of a moose. Another account compares the head to that of a cow. Yet another said it looked like a horse's head with horns that had little knobs on top. That is perhaps what one sad husband saw when he climbed up high on Black Bear Island in Lake Winnipeg after a tiff with his wife. What he thought at first to be a yawl cutting through the waves turned out to be the fore part of a large animal that resembled a sea horse. It too had horns with knobs on the tips. These descriptions are most common, and seem to fit not only other sightings in Lake Winnipeg, but they also apply to a host of other lakes in the country. I have discussed this creature at greater length in the Manipogo chapter.

I will relate a very recent sighting here since it took place where Lake Winnipeg and the Dauphin River meet. It may have a relation to the Manipogos as well, but that is not yet determined. However, since the head of it was clearly seen, that makes this a very rare and significant sighting.

It happened in the summer of 2008, on July 16, as Tony Flett of Fairford was standing on a dock at Dauphin River waiting for a boat to take him across the river to join his family. He had been waiting a long while already, and as he gazed towards the open lake a huge head and neck suddenly rose out of the water a short distance away. It seemed to contemplate the shore, oblivious to the fact that it was not alone. It reached out its neck and nosed some shrubs on the bank before turning towards the lake and diving into the river. It was visible for about five minutes, and during that time Tony studied the creature carefully. It was maroon with a camouflage-like pattern similar to a local fish called a mariah. Its nostrils opened with eyelid-like flaps as it emerged, and its eye appeared as a six-inch-long slit. Where an ear would typically be, a plume of sorts lay against the top of the head, dripping water. The size of the neck Tony compared to the size of his own hefty body, and the head he said was larger than that of a moose's. It never opened its mouth.

When the boat came for him shortly thereafter, he hesitated getting into it, and a lady mentioned that there had been quite a few "snake" sightings that summer.

The maroon color got me all excited since it reminded me of another account, so it began to appear that a composite of this super-serpent was beginning to emerge.

Perhaps 20 years ago, father, son, and girlfriend had gone for a boat ride across Lake St. Martin from Fairford one summer, and they had pulled into shore at German's Beach. Large waves that their boat hadn't created washed around them, causing them to wonder what was going on. They spent about an hour on land before getting back into the boat, and as they exited the bay, they noticed part of a long, maroon object that showed about a foot above the surface, slowly sinking until it was out of sight. It seemed obvious, then, what had caused the big waves earlier.

They had not seen the head, but the spine section of the large back appeared flat. The size of the creature seems remarkably similar to the one Tony saw at the opposite end of the Dauphin River, so if the two sightings possibly showed parts of the same creature, then we can visualize its general appearance. Since there is a correspondence between the color and the size, the chances are good that they are one and the same, but until someone sees the animal from end to end, nothing is definite. It would appear at this juncture that it may be strictly serpentine without any limbs (or flippers), but that is not certain.

I wonder too if this might be one of the large "logs" or "hydro poles" that are so often reported. The girth and length certainly are similar, and if closer observation indicated a maroon color, they might just be a match.

The fearsome head that caused the farmer to streak home many years ago might just belong to this denizen of Lake St. Martin.

Another sighting that recently came to my attention sounds as if it might be related to this maroon monster. Years ago, some Fairford folks were enjoying Hilbre Beach near the south end of Lake St. Martin. The water was calm when the back of a large creature was seen approaching the shore where the people were. When it submerged suddenly, it created large waves that rolled away from it, prompting the adults to hustle the children out of the water.

In addition to the above stories, of course, are the lake monsters that have been mentioned elsewhere under other headings, like the crocodilians, and the long, grey, serpentine "bloodsucker."

Image of a maroon-colored lake monster at Dauphin River sketched by witness Tony Flett.

PUKATAWAGAN MONSTERS

I HAVE ALREADY MENTIONED a number of unusual creatures that were reported in the Pukatawagan area, but the nebulous lake monsters that stirred the community now and then over the years took on more significance in the summer of 2009. That was the summer I attempted to reach one of the caves some residents of this community told me about. In fact, I got the news the day after my failed attempt, as I was wandering throughout the community, still nursing my disappointment.

Apparently someone had captured the monster on video—and my imagination went into overdrive. . . .

The short version of the story is that late in the evening, after I had almost given up hope, a young couple arrived at my abode with the incriminating video still on their little digital camera. Before they even showed it to me, they mentioned something that was significant to them—the eye sockets that were barely visible at the top of the head.

I don't recall being overly impressed with the pictures I saw on the little screen, but it was obvious that something with a sizeable head—assuming it was a head—was swimming across the lake in full view of hundreds of residences facing the shore of Pukatawagan Lake.

It was her baby that got Brenda up before six that morning, but it was already daylight, and as she glanced out her window, she noticed something out on the calm waters. A little later when she looked again, it was closer, so she grabbed her camera and began filming, her baby obviously being part of the process. Then she got some stills as well, one of which shows the vague shapes resembling eye sockets. Not knowing much about photography, I was blissfully confident that enhancement would do wonders for the pictures.

After some negotiating, we agreed to finalize the arrangements for my borrowing the pictures next morning before I caught the train back to Cranberry Portage where my vehicle was waiting for me.

Again my hopes were almost dashed when we left for the long ride to the train station without the pictures. But again, at the tantalizing last minute before I got on the train, a vehicle pulled up, and a very formal agreement was hammered out on a scrap of paper on the hood of the vehicle.

Months later, I was still searching for expertise in interpreting the pictures, and in the meantime, Brenda had again been in the right place at the right time, and saw the head and neck of something else in the lake—only this time she did not have her camera. Others apparently saw it too, so Pukatawagan will continue to be a place I visit regularly.

The creature that Peter Ballantyne saw in the lake at Pukatawagan.

Her baby woke Brenda Dumas early one morning, allowing her to capture this creature on camera.

HORSES?

I BELIEVE THAT THE UNDERWATER MOOSE have been mistaken for horses on numerous occasions because of their un-mooselike appearance, generally. When horses have allegedly been seen in southern lakes under the ice, there is the possibility that the animals were something else. However, some Native accounts hold that these "horses" are thought to have mixed with terrestrial horses in the area, yielding offspring that were untrainable.

From South Indian Lake comes the report from an old man that his parents saw what appeared to be more of a horse than a moose, because it had a tail like a horse. Similarly, a recent report from east of Lake Winnipeg mentions a man shooting a moose that had a tail like a horse.

In the far north, horse tracks were allegedly seen together with caribou tracks.

The flying horses that I hear about are usually referenced in connection with storms, and I tend to lump such accounts together with reports of men with wolf-like heads, or human-like figures that leave moose-like tracks.

LITTLE PEOPLE

ALMOST EVERYWHERE I GO among the mostly Native communities, mention is made of the little people, or *memegweeshig*, also known as the "rock people" because of the rocky areas where they are generally seen. The consensus seems to be that it is a "gift" to have the privilege of seeing these apparitions.

I definitely believe that these beings are a departure from the animals that are represented in the rest of this book, in that they embody a spiritual dimension. They should not, therefore, even be considered alongside all the others that I whole-heartedly believe to be as much flesh and blood animals as the ones we already know all about. I am only treating them here briefly for two reasons. The first, which I have alluded to already, is the unique popularity that they enjoy. The second is to share some observations and perspectives of interest that I have gained over the years.

By far the majority of witnesses seem to be children, as if their size and innocence makes them more compatible with the little people. Not only do some children see them, but they also interact with them in play. And, like in the stories of leprechauns and elves, they seem able to disappear quite quickly.

I have had adults tell me that they saw them passing through doors without opening them. Some men needing aid claim to have had the likes of these appear and provide what was needed. Sometimes they simply appeared, singly or in numbers, curiously investigating the surroundings.

In one community someone saw a host of them climbing all over a newly built bridge as if examining a novelty, but when confronted, all dove into the water.

Native elders with special powers allegedly have been invited into the comfortable interior of their rock abodes and treated hospitably and kindly before being miraculously returned to their homes.

Numerous reports mention seeing the rock people traveling with their little stone canoes, and when disturbed, they tend to drive their crafts into rocks and disappear.

These little people do not seem to have the "mischievous elf" reputation so much as that of curious playmate or wayside helper, although I have heard a number of complaints against them that some were caught taking fish out of people's nets.

That they have a spiritual dimension, I have no doubt, but whence it is, I know not. The other unusual creatures I deal with are, I am certain, none other than the more elusive, covert, and percipient products of an intelligent designer as are the ones we see more often. I believe that they were created for our enjoyment like all the rest, but with that added quality of impressing on mankind the fact that we will never be able to fathom, exhaust, or fully comprehend the limitless creation of God.

A slight deviation from the diminutive rock people who appear to be anywhere from a few inches tall to the more common child-size, are the human-like, bipedal apparitions that some have encountered. For example, one couple, at the beginning of this century, stopped on the road with their car because four of these were right in front of them. The four- to five-foot-tall "people" hopped onto the road and stopped briefly to look at the occupants of the car. They had big eyes, no hair, and smooth, greenish froglike skin with a brown camouflage appearance that made them look as if they had been playing in the mud. Three hopped into the field and formed a huddle, while the fourth went behind the car before jumping the fence. They had the features of humans, and wore only wide strips of rags around the middle.

In contrast, the little people are seen fully clothed, inconspicuously attired. The couple that described their encounter to me seemed embarrassed, and were reluctant to speak about it, as this was obviously a rare and unusual sighting. I can not recall any story quite like this one in my travels, but the description seems vaguely familiar in connection with stories of UFOs, little green men, and such.

WINDIGO/WEETIGO

Since I am violating the scope of this book by including the little people, then I might as well mention the curious Windigo. I wonder sometimes if these beings aren't just the scapegoats who get the blame for anything that is scary and weird. I can still remember hearing the term for the first time in my classroom at Little Grand Rapids from a Grade Five student back in 1966. Everything was a novelty that year—airplanes, canoes, isolation, reservations, Natives, and their culture. We had been talking about something mysterious one day, and the word "Windigo" came out—in the Saulteaux/Ojibwa language.

Since then I have heard a variety of explanations as to what a Windigo is, and perhaps it is just that—a variety of things, depending on people's perceptions.

Mysterious creatures like the Sasquatch have been referred to as Windigos, as have a host of others. But one kind that I am still struggling to get my mind around are the ones that allegedly originated as humans. For one reason or another, they found themselves living in the wild as sub-humans, ostracized by their community. Some are thought to have suffered from a mental illness. Others were banished because of their behaviour. Some lost their minds and turned into Windigos as a result of eating human flesh during times of famine. But I suspect that some, if not all, were driven into the wilderness by the curses of powerful shamans.

I believe we seriously underestimate the abilities and powers that medicine men possessed, more so in the recent past than in the present.

Regardless of what caused these individuals to find themselves on the outside, they were to be feared, for they assumed qualities that were not exactly human, never to return to their former state. For example, they seemed not to be affected by the cold, and yet were cold as ice. And besides having voracious appetites, these cannibalistic beings were commonly known to chew off their own lips, as well as flesh from other parts of their bodies. Some thought that they died once winter was past. Certainly they were something that everyone feared, since they had been known to devastate settlements when they returned at night looking for food.

I heard a number of reports in the north of hunters shooting Windigos, but it was not specified if they did so for self defense or for another reason. Then again, some of these reports may have referred to the Sasquatch.

One shocking account out of the north puts this matter into better perspective. Apparently, less than a century ago, a lone hunter in the wilderness saw another man approaching, so he waved him in for company. He was ignored, however, and as the stranger walked on by, the hunter saw a human arm sticking out from a packing bag. He shot the stranger on the spot with his shotgun, and hastened to fetch others to the scene, where they found other human body parts. The gentleman who shared this story with me concluded by saying, "That was the last time cannibals were here."

For comparison I will mention that a teenage girl who had been lost for about ten years before returning "hairy, naked, and out of her mind" was not referred to as a Weetigo (in Cree territory) in my presence, but I suppose if she had engaged in unusual behaviour, then she might have been considered to be one.

Permit me to stray again, but, a story of a similar nature that dropped into my lap as I traveled through Page, Arizona, in December 2009, is somewhat unusual and not a little intriguing. At a large coal-fired power plant on Navajo land outside of Page, workers on the night shift in October 2009 reported hearing a loud banging of metal objects, so they reported it to security for investigation. The officer on duty, who normally worked only day shifts, drove his truck to the area in question and looked around without success until he saw a swiftly moving dark figure in his mirror. It scaled a set of stairs on the outside of a 45-foot structure, and disappeared somewhere on top.

Nathaniel gave chase, but saw no evidence of the phantom. When he looked down, however, he saw it darting across the yard, away from the buildings. He hurried back down and followed it with the pickup, losing sight of it now and then as he navigated around ponds and other obstacles. In the open, however, he had to drive 35 miles per hour to keep up with it, and, determined to run it down in order to discover what it was, the security guard closed the distance to about five feet. That is when the creature, apparently on two legs, began bounding ahead in about 40-foot leaps, cleared a fence, and disappeared from sight.

When I was curious about its appearance, Nathaniel compared its size to his own 6' 2", 245-pound frame. The weight he said looked similar to his, but its height was about six inches taller. He had only seen it clearly from behind, but the long flowing black hair that covered it from head to toe hid most of its features. It appeared to have shoulders, and yet no arms or wings were visible. The top of the head, with some semblance of ears, reminded him of a dog's head as seen from behind. When Nathaniel returned in daylight to check for tracks, he discovered peanut-shaped footprints that appeared to be moccasin-covered.

I could not resist asking him for his interpretation of the being that he had pursued for about ten minutes, and he candidly gave me an explanation based on a perspective that was relevant to his culture. He felt that the specter was a physical manifestation assumed by a disgruntled or malicious person who perhaps had a grudge against someone working at the plant, or against the plant itself, and was intent on doing harm. His understanding from the elders' stories was that such skinwalkers, or shapeshifter, were usually up to no good. Apparently occurrences such as this one are not nearly as common as they were in the past.

LION

LIONS YOU SAY?

Well, they have manes around their necks, smooth tails with tassels on the end, and look like lions, so what else would people call them?

Mention of them is scattered throughout the north, but not exclusively, as central Manitoba contains stories as well. Some make mention that they were seen in and out of the water, so that is an interesting detail. However, the following account might tend to make us consider that possibility if we could bring ourselves to believe it.

Two women were said to have paddled their canoe near the high cliffs off McBeth Point on the west side of Lake Winnipeg when they found that they weren't getting anywhere. That is when the lady at the back of the canoe noticed a tail holding the canoe back, so she chopped it off. It had been smooth with a tassel on the end.

East of Lake Winnipeg between Berens River and Poplar River is a lake the Natives call "Lion Lake," and holes were seen there where people thought the lions lived. A lady from one of those communities reported seeing two lions somewhere beside the water, while another person had seen a "water lion" run up to the top of a rocky island where it had looked around.

Joshua, from Nelson House, told me of a man who believed these creatures were responsible for eating sled dogs, and he shot one about 60 years ago. Apparently Joshua's own grandfather had shot one too, so he made a rough sketch of the mane and tail for me.

Some other communities where there seemed to be some awareness of lion sightings were Moose Lake and Easterville.

A lady from Queen's University in Kingston, Ontario, who is paddling her way across Canada told me that on one trip she heard a most unusual roar from her tent. It came from the direction of the water, sounding much like the powerful roar of a big lion. The sound was sustained with very little intermission for about ten minutes.

The word "tiger" is sometimes used as well, but perhaps the names are used interchangeably.

It is interesting that lions are known to have lived in North America in the recent past, but the majority of bones belonging to the North American lion have been found in Alaska and in the La Brea tar pits in Los Angeles. The bones show that this animal was considerably larger than the African lion of today. It is believed that they became extinct about 10,000 years ago, but whether or not there is a residual population of them remains to be seen.

TURTLES

TURTLES COME IN A VARIETY OF SIZES and are common to most areas where there is water, but the extraordinary size of some make them worthy of mention. Take, for example, the "sea turtle" that was seen in an inland lake at Little Grand Rapids. The man who described it to me compared its size to a large dining room table. Instead of having legs, it had flippers, and elders were not unfamiliar with it.

The chief, who was a former student of mine, told me that his father-in-law had spoken of a lake south of the community that had trampled muskeg all around it where sea turtles were known to live.

In the Berens River, a four-foot turtle with flippers was seen, and another in some lakes north of Nelson House and in the Churchill River west of Pukatawagan.

A two- to three-foot-diameter turtle was spotted from a ferry in Split Lake, and it was seen snapping at sea gulls. They were not known to live that far north.

Turtles apparently are common in the Assiniboine River, but a man from the Birdtail First Nation told me he saw one that was between four and five feet long. A similar-sized one was spotted at Sioux Valley on top of a beaver's feed pile near the Assiniboine River.

What lives under muskeg and raises the surface vegetation as it moves along? A few hunters have observed this phenomenon, and have been puzzled by it. From the large animals I have heard about, my guesses would include big turtles, big frogs, the giant beaver, and giant snakes.

VICTOR THOMAS

THIS GENTLEMAN IS SIGNIFICANT in my research because he was the first person to present me with a creature other than the giant snakes that I had recently been made aware of from a community neighboring his. He spent a few days in my campground and became fair game for questions that I made a habit of fielding towards anyone living near bodies of water, but he practically ignored my interest in snake stories in favor of what he had seen as a young man.

A sheet of paper quickly showed me what Victor was describing. Their fishing camp had been somewhere on the east shore of Lake Winnipeg, and all fishermen were grounded there one day due to high winds that kept them off the lake. Two unusual creatures were seen swimming under their dock, so, men being men, they fished them out of the water with some kind of hooks and killed them. The adult was about eight feet long, and the juvenile a little more than half that. The men burned as much of the creatures as they could, and buried the remainder.

It appeared that these prehistoric-looking animals were foreign to everyone. And in the more than dozen years I have been in possession of the sketches, I have shown them to hundreds of people. No more than a handful indicated any recognition. What may have been a young one of the species was found dead on the shore of Lake Winnipeg.

Another man, Jean Marie Francois of Nelson House, when he saw the sketches, said he saw two such creatures swimming in Lake Winnipeg near Grand Rapids

STRANGE CREATURES SELDOM SEEN

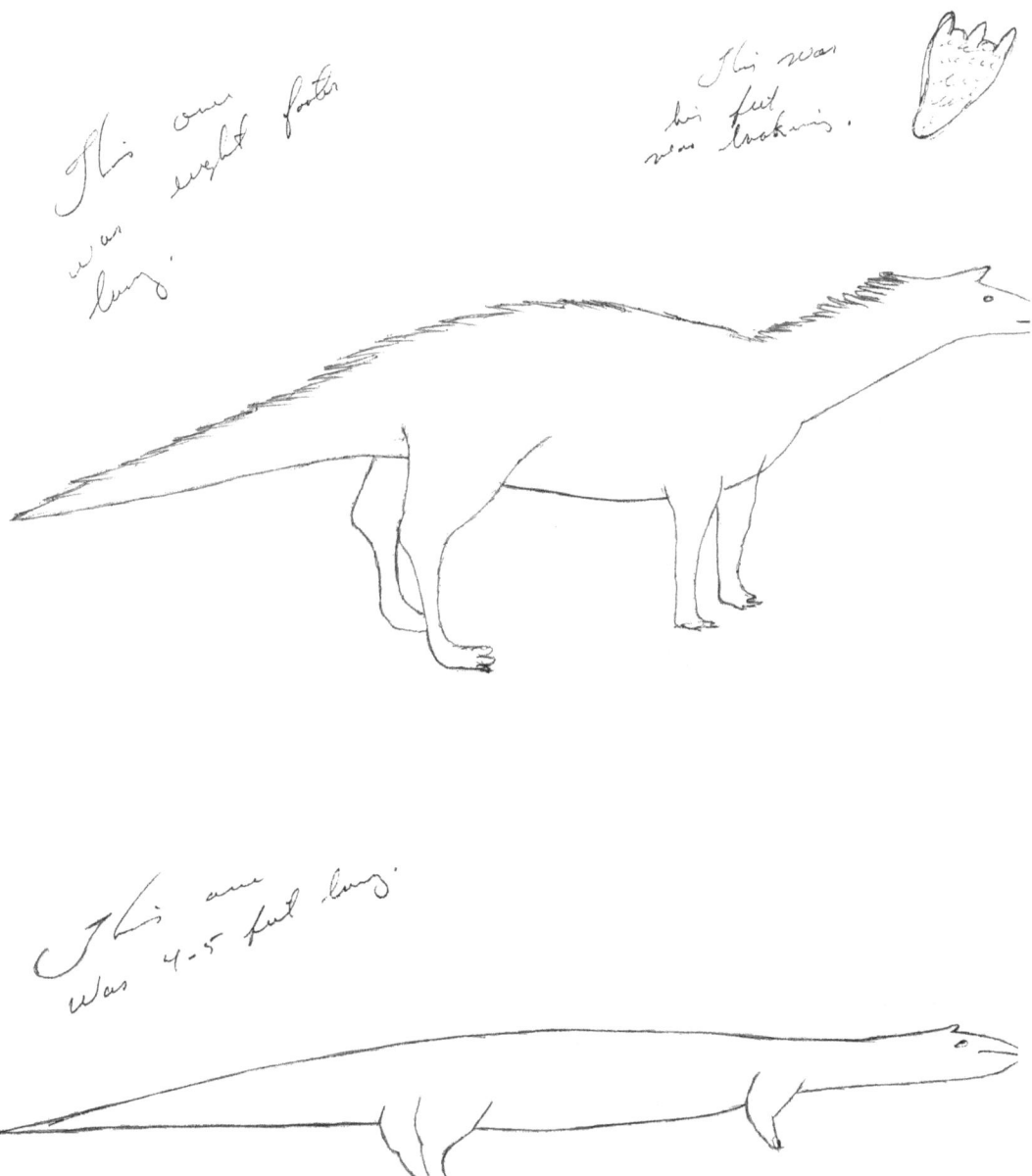

Seen and sketched by Victor Thomas, this adult and juvenile were killed along the shore of Lake Winnipeg.

when he was there firefighting years ago. They swam towards and past each other a number of times, as if in a ritual, with their heads out of the water. This was the same lake as the original sighting, and the account was given to me face to face.

A third sighting may or may not be the same creature, but if it is, then it certainly does add a new dimension to it. The witness said he saw a creature similar to the sketch, also on the shore of Lake Winnipeg, but it sat in a hunched position with its short front legs in the air, much like a squirrel. The tail had been between five and six feet long, and it had jumped into the water "like a loon" when he approached.

I doubt that the two animals are one and the same, but both are unusual for what we know about animals in the region.

JEAN MARIE FRANCOIS

ANOTHER UNUSUAL CREATURE that Jean Marie Francois saw (which I named after him!) was right near his home at Nelson House about ten years ago. He went out a short distance from where he lived with his canoe and spotted something on a small, flat, rocky island. As he paddled towards it, his paddle bumped the canoe enough to wake it up, and he watched it get up and take a few slow steps to the water. It had short black smooth fur like a seal, stood about two feet high, and in length was between four and five feet.

It was not like anything he had seen before in his many years as a hunter and trapper, and he remains puzzled as to what it might be. He told me of another man who had seen something like it, and, thinking it to be a bear, had paddled after it, only to see it submerge. Others had seen it too, but no one ever caught or shot one to find out what it really was.

The size sounds similar to the water dogs, but that is where the similarities end. The water dogs were brown, dog-footed, and fast moving. This creature was black, slow-moving, and had large, extended feet. One couple whose house looks out on the bay where that little island is visible tell me that it is common to see something lying on the rock some summers. So, why don't we . . . ?

RABBITS?

WHAT ABOUT SUPER-SIZED RABBITS? I've heard mention of a rare big one, but how about a seven footer? A husband and wife were standing at the end of their field in northern Alberta looking down into a ravine. There they saw what appeared to be a huge, tan-colored rabbit that even hopped like one and had ears similar to one.

STRANGE CREATURES SELDOM SEEN

The creature seen and sculpted by Jean Marie Francois of Nelson House.

THE MORDEN DOGCAT

AROUND THE TURN OF THE MILLENNIUM there was a sizeable creature seen in southwestern Manitoba that still has some of the rural residents puzzled. It looked like a dog and loped like a dog, but something about its head and its demeanor suggested otherwise. Folks living or driving south of Morden near the Pembina Hills saw it off and on over a period of several years, and then sightings ceased altogether.

It was seen in broad daylight by some, and in the headlights by others, in yards, and crossing roads. It had a somewhat shaggy dark brown coat on about a 75-pound frame, but its distinctive feature was the compacted muzzle that gave its head more of a cat-like appearance. It was just not like any other creature that anyone had ever seen. One farmer saw it cross his field and return with a rabbit in its mouth. He thought it had a lynx-like appearance, but a hunter who was familiar with lynx and saw it cross the road just ahead of him in daylight did not recognize it as such.

Dog-like creature seen near Easterville.

BIG CAT?

ON NOVEMBER 11TH OF 2007, Mike Bear of the Sandy Bay First Nation in northeastern Saskatchewan shot and killed two cats out of a group of about ten a few miles from the community when he was out hunting with a cousin and their dogs. The creatures, which appeared to be super-sized domestic cats, ran into the bush and climbed trees when chased by the dogs, and were even seen jumping from tree to tree. A dog that followed them into the bush returned badly scratched.

GIANT SPIDERS

A MAN FROM BERENS RIVER told me that he was going to set a trap for otter in a big crack in a rock, but changed his mind when he saw a huge, black spider-like creature there that he thought might attack him. Its long legs were thicker than a man's thumb, and its eyes were as big as quarters. The body size was about a foot in diameter—not counting the legs, and it was 'hairy like a monkey." Its foot-long tail was about an inch thick. When it stood up on its long legs, it was about a foot high.

I heard only one other rumor regarding such spiders in all my travels.

Never-Before-Seen Species of Cat Killed Near Sandy Bay

A strange species of cat, never-before-seen, is shot by Mike Bear of Sandy Bay on November 13 at Burntwood Lake, three miles west of the community. Mike was with a younger cousin when they saw the cats that numbered about 10 in a group, all the same size. Mike shot and killed two of them with a .22-calibre rifle.

Everyone who has heard of this happening in the Sandy Bay-Pelican Narrows area is amazed of hearing about, and seeing photographs of, this mysterious species of cat. Charlie O. Morin of Sandy Bay opines that it may be a cross between a domestic house cat and maybe a lynx. It has five claws each on the forelegs and four claws each on the hind legs, and has yellow eyes. One dog that chased the other cats into the brush returned wounded with cuts and scratches.

The dead cat appearing in the photos is smaller than some of the others in the group. They are capable of climbing, and reportedly were seen jumping from tree to tree after being chased up one by dogs accompanying Mike Bear and his cousin.

Bear shot the cat, appearing in the photos, five times with his rifle, and only one bullet went all the way through its body, so they are apparently strong and muscular. Kwayask or iswwak itinawak, Pakwacha!

Sandy Bay feline

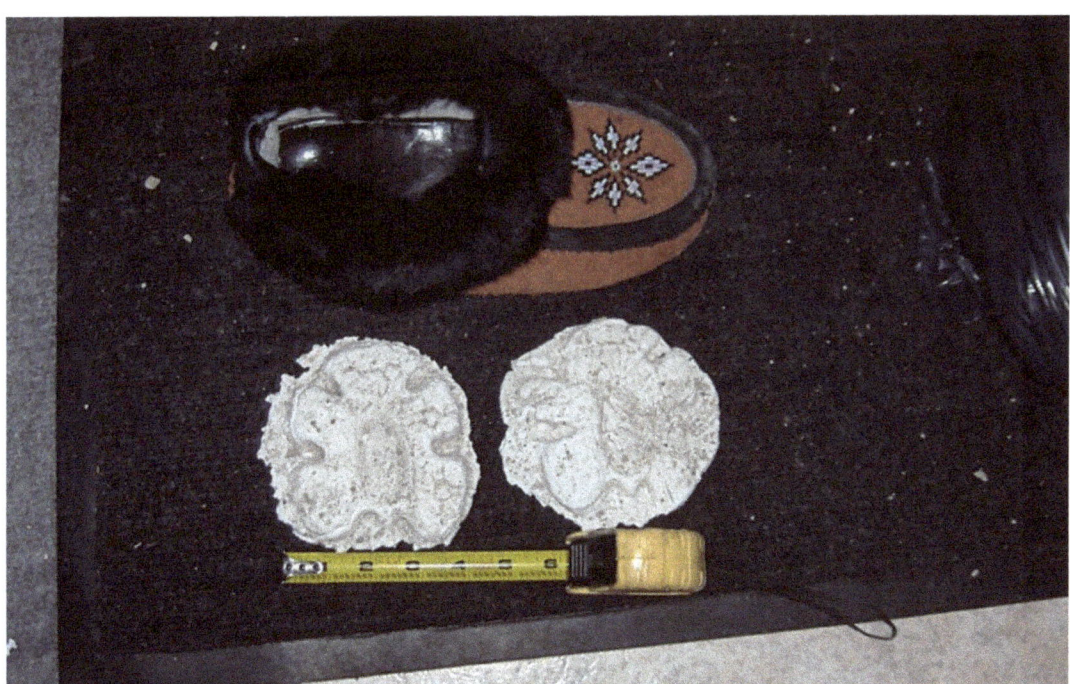

A set of footprints seen in Fairford campground one morning after a rain.

WHAT'S IT?

How about a blue—yes, blue—overgrown mouse? Cut a mink in half and you have its size. It ran across the beam of light from a machine working on the Wuskwatim Dam project—in winter. It had no bounce in its gait, but ran smoothly like some mice do. And it was blue!

I believe John because he has already given me other stories that I found believable—like the two sets of Sasquatch stories from west of Nelson House, and the short-legged moose on Highway 6. Yes, I know what you're thinking. But believe me . . . believe him.

And then there is the fish with three eyes, one with two mouths, another with legs, and still another with the face and teeth of a dog.

MISPLACED ANIMALS

Let's go just a little off topic and see what folks have told me about animals that aren't exactly where they are expected to be. Like the young penguin and seal at Pukatawagan, the grizzly near Cross Lake, the polar bear shot at Oxford House and another seen at Pikwitonei . . . and the walruses near God's Lake Narrows that the ancestors spoke about.

Then there are the various critters that are working their way north. Wood-ticks are inching past Grand Rapids. Deer, skunks, cougars and vultures are among others that are seen in places where they were not known before. At Shamattawa folks mentioned the appearance of pelicans, coyotes, and skunks.

One lady from Wabowden commented that she had grown up with long spells of very cold winter and very hot summers, and that pattern, among other things, had changed considerably during her lifetime. As I write this, however, in the winter of 2008-2009, the weather regressed reminiscent of the olden days, giving us phenomenally cold weather.

PICTOGRAPHS AND PETROFORMS

Although I have not seen any pictographs or petroforms myself that portrayed any of the above-mentioned unusual creatures, I have been told of quite a few that seemed to indicate larger than ordinary species of animals like snakes, birds, or beaver. They, together with the oral histories and present-day sightings, seem to indicate unusual creatures that were or are present in Canada.

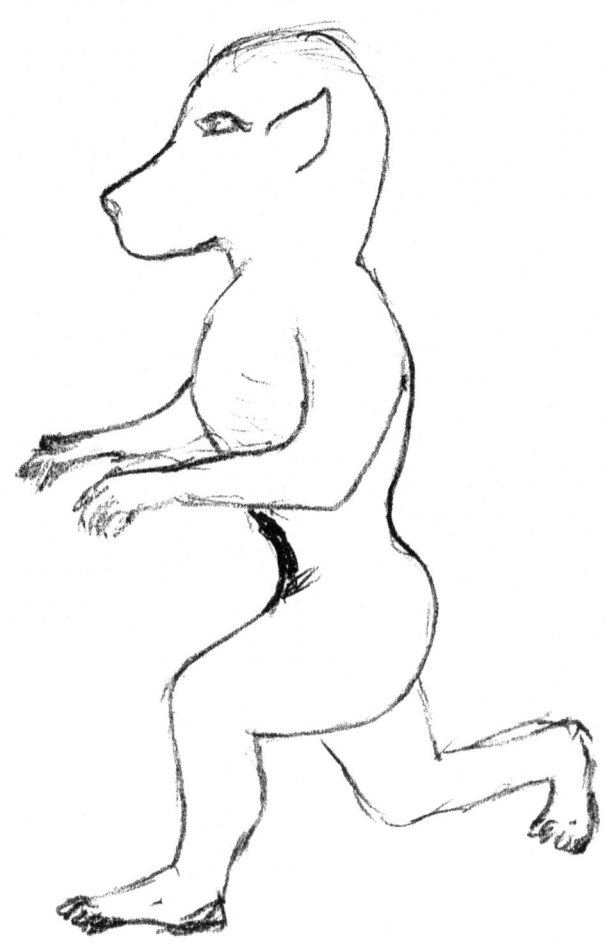

One last mystery: a biped seen by biologist
Pat Rousseau and partner in northern Quebec.

Coachwhip Publications
CoachwhipBooks.com

The Historical Bigfoot
Chad Arment

ISBN 1930585306

Available from your favorite online bookstore.

Coachwhip Publications

Also Available

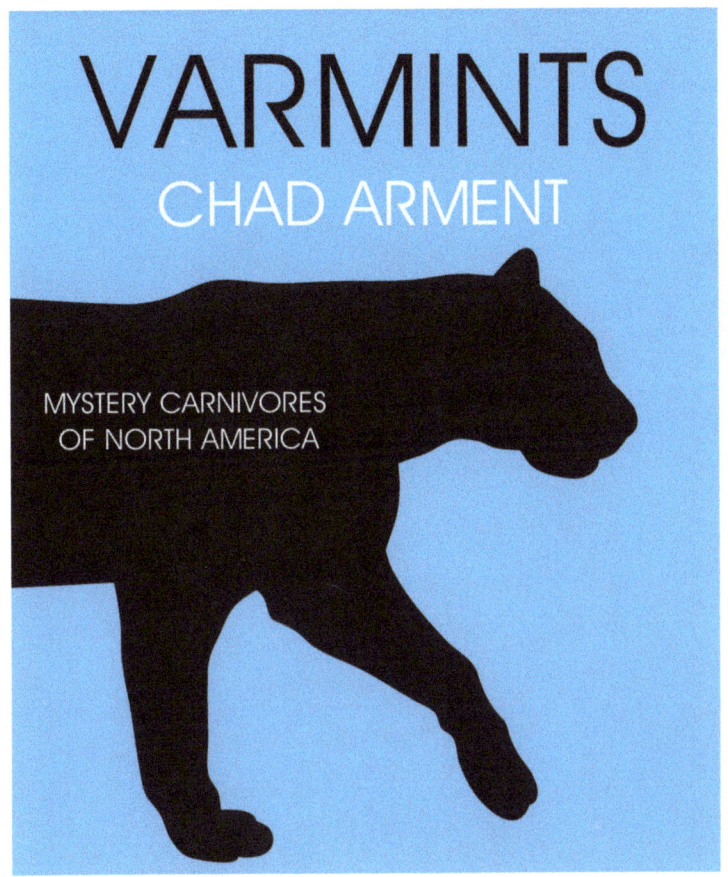

Varmints: Mystery Carnivores of North America
Chad Arment

ISBN 1616460199

Available from your favorite online bookstore.

Coachwhip Publications
CoachwhipBooks.com

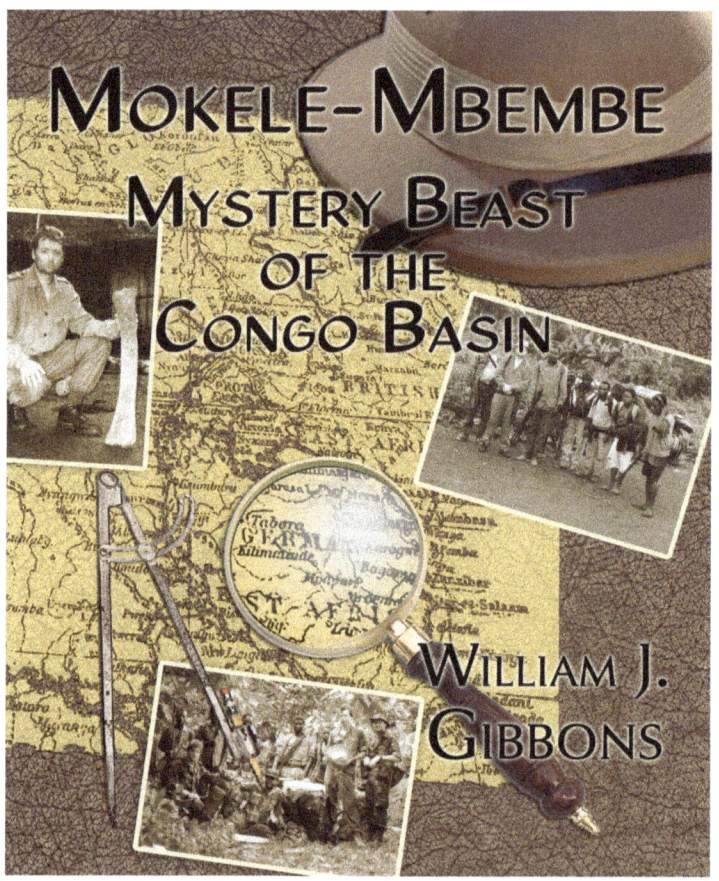

Mokele-Mbembe: Mystery Beast of the Congo Basin
William J. Gibbons

ISBN 1616460105

Available from your favorite online bookstore.

Coachwhip Publications
Also Available

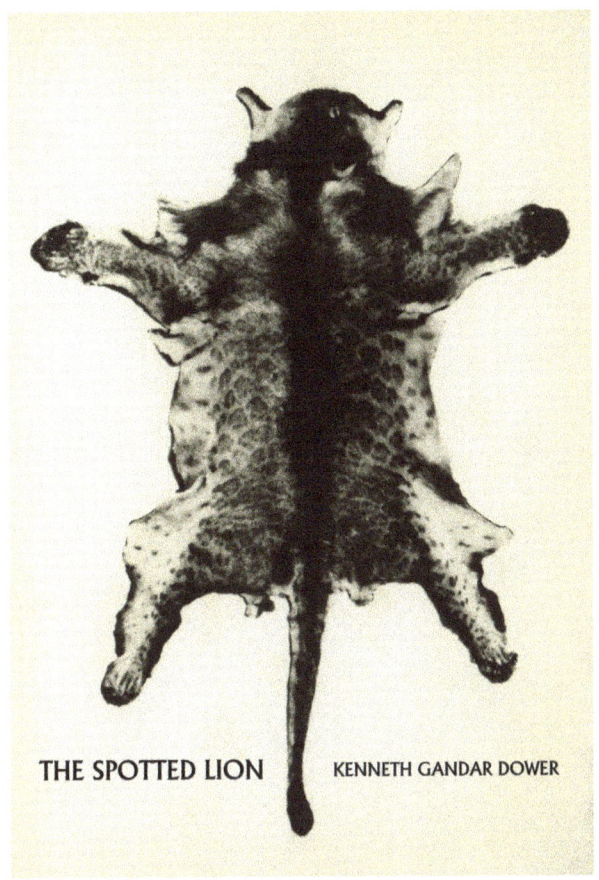

The Spotted Lion
Kenneth Gandar-Dower

ISBN 1616460717

This title is available from BookDepository.com.

www.ingramcontent.com/pod-product-compliance
Lightning Source LLC
Chambersburg PA
CBHW040732020526
44112CB00059B/2946